Philosophies of Mathematics

MADELEINE CLARK WALLACE LIBRARY
WHEATON COLLEGE
NORTON, MASSACHUSETTS

George and Helen MacGregor Paul '27
Library Fund

Philosophies of Mathematics

Alexander George and
Daniel J. Velleman

BLACKWELL
Publishers

Copyright © Alexander George and Daniel J. Velleman 2002

The right of Alexander George and Daniel J. Velleman to be identified as authors of this work has been asserted in accordance with the Copyright, Designs and Patents Act 1988.

First published 2002

2 4 6 8 10 9 7 5 3 1

Blackwell Publishers Inc.
350 Main Street
Malden, Massachusetts 02148
USA

Blackwell Publishers Ltd
108 Cowley Road
Oxford OX4 1JF
UK

All rights reserved. Except for the quotation of short passages for the purposes of criticism and review, no part of this publication may be reproduced, stored in a retrieval system, or transmitted, in any form or by any means, electronic, mechanical, photocopying, recording, or otherwise, without the prior permission of the publisher.

Except in the United States of America, this book is sold subject to the condition that it shall not, by way of trade or otherwise, be lent, resold, hired out, or otherwise circulated without the publisher's prior consent in any form of binding or cover other than that in which it is published and without a similar condition including this condition being imposed on the subsequent purchaser.

Library of Congress Cataloging-in-Publication Data

George, Alexander.
    Philosophies of mathematics / Alexander George and Daniel J. Velleman.
        p. cm.
    Includes bibliographical references and index.
    ISBN 0–631–19543–2 (hardback : acid-free paper)—ISBN 0–631–19544–0
    (paperback : acid-free paper)
        1. Mathematics—Philosophy. I. Velleman, Daniel J. II. Title.

    QA8.4 .G46 2002
    510′.1—dc21

                                                                    2001037590

British Library Cataloguing in Publication Data

A CIP catalogue record for this book is available from the British Library.

Typeset in 10/12.5pt Sabon
by Kolam Information Services, Pondicherry, India.
Printed in Great Britain by T.J. International, Padstow, Cornwall

This book is printed on acid-free paper.

Contents

Preface

Philosophy of mathematics might at first appear a rather unlikely marriage. Philosophy is sometimes viewed as the repairman of the sciences, and mathematics, it so obviously seems, is in no need of fixing. Mathematics enjoys the unique reputation of being the acme of clarity and certainty. What has philosophy to teach it? Have philosophers not instead regularly sought to put their own house in order by emulating mathematics with all its sterling virtues?

It is true that while philosophers perennially debate about the nature of philosophy, there seems to be little comparable disagreement among mathematicians. But that is only because mathematicians are usually busy doing mathematics rather than thinking about what they are doing. It is notable that whenever mathematicians do pause to have a discussion about this, very little agreement emerges. The last such pause of any historical consequence began toward the end of the nineteenth century and continued until the 1930s. During that time, mathematicians mounted a sustained effort to clarify the nature of their enterprise. It led to considerable discord, even enmity, and yielded fascinating and fruitful work of both a mathematical and a philosophical nature; it was one of the most exhilarating intellectual adventures of the twentieth century, pursued at an extraordinarily high level of acuity and imagination. The debate gradually subsided, and relative calm about these matters is now again the norm within the mathematical community. Yet, while the central philosophical issues that divided the protagonists remain, we have been left a rich treasure of ideas, philosophical and mathematical, with which to continue the inquiry.

The legacy principally consists of three original and finely articulated programs that seek to view mathematics in the proper light: logicism, intuitionism, and finitism. Each is notable for its symbiotic melding together of philosophical vision and mathematical work: the philosophical ideas are given their substance by specific mathematical developments, which are in turn given their point by philosophical reflection.

These projects are in turn intimately related to one another, and there is a sense in which none of them can be fully appreciated without being viewed against the background of the others. They are also integrally connected to the most important contributions in the philosophy of mathematics that have come in their wake: for contemporary work is best evaluated against a backdrop of understanding that encompasses these three great historical attempts to tame the phenomenon of mathematics.

As teachers, we find this fascinating material of particular interest because it is thoroughly accessible to any undergraduate with a background in logic. No sophisticated philosophical or mathematical training is required; just familiarity with quantificational logic of the kind that can be gleaned from any introductory course or textbook, and a taste for the problems and approaches involved. We were consequently eager to bring this exciting intellectual fare to Amherst College, which has been blessed with a steady stream of students with the right appetites. To our surprise, we found that no book existed that offered a thorough introduction to the philosophical and mathematical details of this chapter of intellectual history. There are advanced treatments to be sure; and collections of seminal texts, both philosophical and mathematical. But we have found few if any contemporary works that introduce and carefully develop the philosophies, the mathematical projects, and their complex interconnections. We hope we have succeeded in writing such a book; at any rate, it is what we aimed to produce.

It has been observed that one of the pressures leading to the clarification of the foundations of the calculus was the fact that mathematicians increasingly found themselves in the awkward position of having to teach material they did not fully understand. We have found likewise that actively teaching about these three philosophies of mathematics has forced us to deepen our understanding of, and develop our judgments about, both the philosophical and the technical material we shall present. Of course, our reading of secondary literature has been useful as well. That said, for expository reasons we have chosen not to engage explicitly with any alternative interpretations of these philosophies or to enter overtly into any scholarly debates. There is certainly room for treatments that are more ecumenical in nature or, instead, for those that are more openly sectarian. But that is not the kind of book we have tried to write.

Finally, it bears remarking that this book does not view the central mathematical and philosophical projects through the eyes of the historian. No sustained attempt has been made to trace the historical development of these ideas, nor do we even seek always to remain historically faithful in their articulation. Our guiding concern has been to present central positions and arguments in the clearest and strongest form, and with a view to highlighting their conceptual interest; if we judged that an anachronism would aid in that attempt, we have not scrupled to employ it. Still, we trust that what follows gives a faithful and fundamental rendering of the deep ideas in this debate, and that it captures both the spirit and the texture of these enduring contributions to our thought.

We are grateful to Amherst College and the National Endowment for the Humanities for their support. We also thank our teachers and students, and our families and friends.

Alexander George
Daniel J. Velleman
May 2001

1

Introduction

The links between philosophy and mathematics are ancient and complex. The two disciplines are in a sense coeval: for both, the Ancient Greeks were the first to introduce systematicity, rigor, and the centrality of justification to their practice. Indeed, Plato (428–348(7) BCE) had it inscribed on the gates of his Academy that no one should enter who knew no mathematics. There have since been few major philosophers in the Western tradition who have not labored mightily to understand the phenomenon of mathematics.

Yet it might at first seem a surprise that philosophy, which often concerns itself with fundamental issues arising out of everyday life, should focus so on perhaps the most abstract discipline, one whose subject matter and methods seem, as we shall soon see, so removed from ordinary experience. While there is surely no one explanation for this, and certainly no one answer that can be summarized without some injustice, we might point to at least the following.

Our knowledge of mathematical facts is often thought to be justified by pure ratiocination, perceptual observation via any of our five sensory modalities being neither necessary nor even relevant. Because mathematical knowledge seems to be unconstrained by experience, and so to enter the world touched only by the hand of reflection, analysis of it promises to reveal something important about rational thought itself. Since the nature of rationality has always gripped the philosophical imagination, it is perhaps no mystery after all that philosophers regularly direct their attention to mathematics. Mathematics is the purest product of conceptual thought, which is a feature of human life that both pervasively structures it and sets it apart from all else.

Other reasons for philosophical interest in the nature of mathematics will emerge as we proceed. It might be useful to pause, however, and to ask what distinguishes a philosophical interest in mathematics from other kinds of concern. Although the question of what counts as such an interest admits of no briefer or simpler an answer than does that about what philosophy itself is, we can approach it helpfully, if indirectly, by first describing the kinds of study of mathematics we are *not* interested in here.

To begin with, we shall not engage in a *historical* investigation into the development of mathematics. The etiology of mathematical ideas, however interesting, is not something whose study promises to reveal much about the structure of thought: for the most part, the origin and development of mathematical ideas are simply far too determined by extraneous influences. The same holds for a *sociological* inquiry into the role of mathematics and mathematicians in our society, the forces that shape research interests and structure professional activity, and so on: again, these aspects of the doing of mathematics bear the imprint of countless varied factors and so it is difficult, if not impossible, to distill from such inquiries information pertinent to central philosophical concerns.

Since these concerns focus on the nature of rational thought, it might appear especially strange that we are likewise not going to pursue any kind of *psychological* inquiry into mathematical thinking or development. Such research often takes for granted a conception of thought, usually undeveloped, and asks such questions as "What brain, or neural activity, or cognitive architecture makes mathematical thought possible?" or "What kind of environment is needed to facilitate the development of the capacity for such thought?" Again, while of great interest, such studies focus on phenomena that are really extraneous to the nature of mathematical thought itself. Indeed, to repeat, they often proceed without a developed account of what such thoughts are, and they concentrate rather on the neural states that somehow carry thought, or on the environmental or genetic factors that make those states realizable. Philosophers, by contrast, are interested in the nature of those thoughts themselves, in the content carried by the neural vehicles (if that is indeed the right picture). A philosophical study of mind is interested in an analysis of the thoughts that the workings of mind give one access to, but not in an account of the conditions or the mechanisms – environmental, genetic, neurophysiological – to whose influence and operation we owe our access to such thoughts.

We should add, finally, that a philosophical inquiry into mathematics differs from a *mathematical* one. (It is interesting to note that perhaps the only discipline other than philosophy so clearly amenable to such self-analysis, so fitted to take its own methods and problems as a focus of inquiry, is mathematics.) What precisely the application of mathematics to itself might consist in is something that we shall take up later. And although we shall then find this self-application of great interest, its import is not philosophically transparent. Rather, as we shall see, it provides data for a satisfying philosophical analysis, but it cannot substitute for that analysis.

Let us turn now to a more positive characterization of a philosophical approach to mathematics. It will be helpful to focus on instances of actual mathematics, so let us consider a few theorems and their proofs, and then survey the kinds of typically philosophical issues they raise. The first two both involve the distinction between rational and irrational numbers. A *rational* number is one that can be expressed as a fraction; for example, 3/5, −19/12, and 8/1 are all rational numbers. An *irrational* number − π, for instance − is one that cannot be expressed as such a fraction. The rationals and irrationals together make up the *real* numbers. Our first theorem dates from Ancient Greece; it is usually atttributed to a member of the school of Pythagoras, although precisely who first proved it is not known. The English mathematician G. H. Hardy (1877–1947) called it a theorem "of the highest class. [It] is as fresh and significant as when it was discovered − two thousand years have not written a wrinkle" on it.[1]

THEOREM 1.1. $\sqrt{2}$ *is irrational.*

Proof. By *reductio ad absurdum.* Assume that $\sqrt{2} = a/b$, for some integers a, b. By reducing the fraction a/b to lowest terms if necessary, we may assume that a and b have no common factor. Then $2 = (a/b)^2$. That is, $2 = a^2/b^2$. Hence, $2b^2 = a^2$. But then a^2 is even. So, since the square of an odd number is always odd, a is even; that is, $a = 2c$, for some c, and $a^2 = 4c^2$. Substituting, we get that $2b^2 = 4c^2$; that is, $b^2 = 2c^2$. Hence, b is even. But this contradicts our assumption that a and b have no common factor (since if they were both even, they would have a common factor of 2). Therefore, $\sqrt{2} \neq (a/b)$, for any integers a, b; that is, $\sqrt{2}$ is irrational. ■

This demonstrates that not all magnitudes − in particular, not the length of the hypotenuse of a right-angled triangle of unit base and height − can be treated by the theory of numerical proportion upon which the mathematics of Ancient Greece was based. It must have thus constituted something of a revolution.

Our second example is of more recent vintage.[2]

---

[1]   Hardy (1967, 92).
[2]   The earliest published reference known to us is Jarden (1953).

THEOREM 1.2.   *There are irrational numbers a, b such that a^b is rational.*

Proof. By argument by cases. Either $\sqrt{2}^{\sqrt{2}}$ is rational or it is not.

 Case 1: $\sqrt{2}^{\sqrt{2}}$ is rational. By Theorem 1.1, we know that $\sqrt{2}$ is irrational; so let $a = b = \sqrt{2}$. Then a, b are irrational and a^b rational, as required.

 Case 2: $\sqrt{2}^{\sqrt{2}}$ is not rational. Let $a = \sqrt{2}^{\sqrt{2}}$ and let $b = \sqrt{2}$. Then $a^b = (\sqrt{2}^{\sqrt{2}})^{\sqrt{2}} = \sqrt{2}^{\sqrt{2}\cdot\sqrt{2}} = \sqrt{2}^2 = 2$, which is rational.
In either case, then, the theorem is shown to be true.   ∎

 Finally, we give an example of a proof that uses *mathematical induction*. Mathematical induction is an important method for proving general statements about the natural numbers: the numbers 0, 1, 2, and so on. To prove a statement of the form "For all natural numbers n, $P(n)$" by mathematical induction, one first proves that $P(0)$ is true (the *base case* of the induction), and one then proves that for every natural number n, if $P(n)$ is true, then so is $P(n + 1)$ (the *induction step*). The assumption in the induction step that $P(n)$ is true is called the *inductive hypothesis*. To see why this establishes the desired conclusion, note that according to the induction step with $n = 0$, if $P(0)$ is true then $P(1)$ is true. But by the base case, $P(0)$ *is* true, so it follows that $P(1)$ is also true. Applying the induction step with $n = 1$, we see that if $P(1)$ is true then $P(2)$ is true; since we have just shown that $P(1)$ is true, $P(2)$ is also true. Continuing in this way, we can establish $P(n)$ for every natural number n. In other words, the statement "For all natural numbers n, $P(n)$" is true.

THEOREM 1.3.   *For every natural number n, $0 + 1 + 2 + \ldots + n = n(n + 1)/2$.*

Proof. By mathematical induction.

 Base case: When $n = 0$, both sides of the equation are equal to 0.

 Induction step: Suppose that $0 + 1 + 2 + \ldots + n = n(n + 1)/2$. Then

$$0 + 1 + 2 + \ldots + (n + 1) = (0 + 1 + 2 + \ldots + n) + (n + 1)$$

$$= \frac{n(n + 1)}{2} + (n + 1) \quad \text{(by the inductive}$$

$$\text{hypothesis)}$$

$$= \frac{n(n+1) + 2(n+1)}{2}$$
$$= \frac{(n+1)((n+1)+1)}{2}. \qquad \blacksquare$$

These examples are in a way paradigmatic mathematics. To be sure, mathematics is filled with proofs that are much longer and more complicated, and with theorems that involve concepts far more intricate than those appearing above. But most philosophical questions about mathematics can already be raised with regard to such simple examples. We shall briefly examine a few of these in turn.

To begin with, note that (assuming you had not seen these proofs before) you now know three more truths than you did a few moments ago. How did you acquire this knowledge? One way you did *not* acquire it is through observation. It is true that you used your eyes to read the sentences of the proofs. But this was by no means necessary. You could, after all, have thought up the proofs yourself (as did those who first arrived at them). And anyway, seeing that the ink is arrayed just so on the page is not what shows you that the theorems are true, for the theorems are not about the disposition of ink on paper.

In this connection, philosophers often distinguish between *a priori* and *a posteriori* knowledge. The latter kind of knowledge requires sensory experience for its justification, whereas the former requires none. Mathematical truths, unlike, say, truths about the natural world, are known *a priori*. You now know that $\sqrt{2}$ is irrational, not on the basis of any measurements or observations, but rather on the basis of pure reflection. This is quite unlike your knowledge that this book weighs less than you do, which is *a posteriori* in that it cannot possibly be justified without some kind of recourse to observation.

To be sure, it might be the case that sensory experience is required in order for us to acquire the language by means of which we can understand the thoughts involved in the above theorems and their proofs. But we must distinguish between what is needed in order for one to grasp the content of these theorems and what is relevant to their justification. The present distinction between *a priori* and *a posteriori* knowledge pertains to the latter only. For example, there is a sense in which you would not have known that Theorem 1.1 holds had (say) page 3 been missing from this book (assuming, still, that you learned it here for the first time); sense experience, in particular your sensory interaction with the appropriate page, is relevant to your *acquisition* of the knowledge that this theorem obtains. The physical page and its properties are not, however, relevant to your *justification* for believing that Theorem 1.1 is true. By contrast, the fact that any fraction can be reduced to lowest terms *is* something upon which the justification of Theorem 1.1 depends. One might express this by saying that, should the claim about reducing fractions to lowest terms not hold, then we would not be said to know that Theorem 1.1 is true. But one should not assimilate this situation to that arising from the missing page. The latter leads to an impediment to our acquisition of knowledge, whereas the former presents an obstacle to our justification. To determine whether a claim is known *a priori* or not, we must examine the kind of evidence upon which the justification of the claim rests; in particular, we must determine whether any of it is based on sensory experience. Theorem 1.1, for example, is an instance of *a priori* knowledge because, in spite of the fact that sensory experience is required in order to come to know it, its justification does not depend on any claims known through experience. The same holds for Theorems 1.2 and 1.3.

The above theorems, then, are known *a priori*. Now this feature of mathematical knowledge, that its justification makes no reference to facts known through observation, appears at first glance to clash with *empiricism*, a most influential and long-lived doctrine according to which all our knowledge is ultimately based on our sensory experience. Of course, if "based on" means merely "would not be possible without," then empiricism cannot be gainsaid any more than other patent truths. So understood, it is not quite a tautology, for there is perhaps nothing absurd in supposing that a being could come to acquire a language and knowledge on the basis of no experience at all, but even the most cursory examination of how it is with humans will show one that it is true. Consequently, if empiricism is to rise above the more than obvious, then "based on" must be interpreted otherwise.

In fact, the doctrine is typically interpreted more strongly to hold that all knowledge is ultimately justified on the basis of sensory experience. And it is with this familiar interpretation of the doctrine that the *a priori* nature of mathematical knowledge is in conflict. The clash is a serious one in so far as empiricism is a plausible doctrine. And it is not hard to see why many have found it such. For much of our knowledge does seem to be rationally based on observations of the world. It might even appear difficult to see how one could have knowledge of the external world except through its rationally impinging upon one in some way. Since our only channels of information are the five sensory modalities, it seems that all knowledge of the world must ultimately be justified by data that are transmitted through them. And yet knowledge of the above three theorems is not.

One way of alleviating the present pressure is to reject the view that these theorems are about the world. Empiricism, after all, appears to be motivated by considering how knowledge of the external world could be justified. Perhaps, then, the entire conflict can be defused by denying that mathematics tells us anything about the natural world. But if it is not about the natural world, then what *is* the subject matter of its claims? One very powerful answer is that mathematics concerns a world that exists quite as independently of us as does the natural world, only one that is not located anywhere in space or time. Hardy eloquently describes this doctrine as follows:

> By physical reality I mean the material world, the world of day and night, earthquakes and eclipses, the world which physical science tries to describe.
>
> I hardly suppose that, up to this point, any reader is likely to find trouble with my language, but now I am near to more difficult ground. For me, and I suppose for most mathematicians, there is another reality, which I will call "mathematical reality." [...] I believe that mathematical reality lies outside us, that our function is to discover or *observe* it, and that the theorems which we prove, and which we describe grandiloquently as our "creations," are simply our notes of our observations.[3]

This view of mathematics is often called *platonism*.

Platonism, as just described, in fact involves two distinct ways of characterizing a realm of reality. The first is *ontological* in nature, that is, it proceeds by describing the kind of entities that inhabit the particular realm in question. Mathematical platonists usually insist that mathematical entities are abstract in not being spatiotemporally located and thus in not having any causal powers. If we think about this aspect of platonism in connection with our three theorems, we cannot but acknowledge its plausibility. For instance, it would be quite odd for someone to think that $\sqrt{2}$ is actually a locatable and datable entity. The request for its location or for the time it came into existence admits of no answer that does not confuse a vehicle for referring to that number (for example, ink marks on paper, brain states of a human) with the number itself. Likewise, it seems implausible that the infinitude of natural numbers should remain an open question until physics determines whether the universe is infinite; but if we insist on identifying each of the infinitely many natural numbers with a distinct spatiotemporally located bit of matter, then this is precisely what we shall have to accept.

Platonism provides a second way of characterizing the realm of mathematics, one that we might call *doxastic* in that it describes the relation between, on the one hand, truths about the realm in question and, on the

---

[3]   Hardy (1967, 122–4).

other, what we believe. Along this dimension, platonism insists that mathematics is mind-independent, in the sense that whether a mathematical statement holds is quite independent of what we think. We can imagine certain realms in which the beliefs of observers in effect settle what is true and what is not. But mathematics, according to the platonist, is not like this: the truth or falsity of a mathematical claim is not determined by what anyone believes about its truth value. This, too, is a plausible position with regard to the theorems above. For instance, the square root of 2 is irrational regardless of whether anyone believes or wants it to be; indeed, its irrationality is not contingent on anyone's having beliefs about it at all. This result obtains, Hardy insisted, "not because we think so, or because our minds are shaped in one way rather than another, but *because it is so*, because mathematical reality is built that way."[4]

There are other senses of mind-independence with which this last should not be confused. One might claim that mathematics is mind-independent in that the entities that make up its subject matter are distinct from items of the mind. (Whether this is entailed by the ontological component of platonism depends on whether mental items are themselves spatiotemporally located.) Or one might hold that mathematical entities are mind-independent in that their existence does not presuppose the existence of minds. (The latter conception of mind-independence clearly entails the former. But not conversely: for someone could consistently maintain that just as an artifact, although an object distinct from the artisan, is dependent for its existence on the latter's activities, so mathematical entities, although not themselves mental, would nevertheless not exist were it not for minds.) We shall return to some of these notions later, but for now let us simply observe that they are both to be distinguished from the claim of mind-independence that we are taking to be part of platonism. For even if mathematical entities are identical to mental items, one can still maintain that truths about these entities hold regardless of our beliefs about them: there is no reason why facts about our minds must be in principle accessible to us. We might note, finally, that a generalization of this last observation shows that the two features of platonism we have isolated are distinct: that the truths of mathematics hold regardless of our recognition that they do fails to settle the ontological nature of the entities these truths concern.

We first encountered platonism as a way of viewing mathematics that defuses its threat to empiricism. But does it really succeed in this? One might well wonder whether talk of abstract entities is less a solution to

---

[4]   Hardy (1967, 130).

the empiricist's problem of how *a priori* knowledge is possible than it is a label for the problem. For it remains to be explained how we, creatures located in space and time, can acquire information about entities that are not so located. Because these latter are causally impotent and so cannot affect us in any way, it seems a mystery how we could come to have knowledge about them. To say that thought happens to permit direct access to this independently existing, though causally quarantined, world might seem, again, less a solution than a colorful description of the originally puzzling situation.

In addition, even if one does grant that mathematics is about some nonphysical reality, the fact remains that mathematical knowledge is very *applicable* to physical reality. Mathematics, from the humblest of arithmetical calculations to the most recondite of theories, finds itself useful in the description and prediction of natural phenomena. Physicists are of course particularly aware of this. Albert Einstein (1879–1955) wrote of "an enigma [that] presents itself which in all ages has agitated inquiring minds. How can it be that mathematics, being after all a product of human thought which is independent of experience, is so admirably appropriate to the objects of reality? Is human reason, then, without experience, merely by taking thought, able to fathom the properties of real things?"[5] As the American physicist Steven Weinberg (1933–) put it, "It is positively spooky how the physicist finds the mathematician has been there before him or her."[6] Johannes Kepler (1571–1630), the German astronomer and mathematician, likewise impressed by this fact, suggested that "God himself was too kind to remain idle, and began to play the game of signatures, signing his likeness into the world; therefore I chance to think that all nature and the graceful sky are symbolized in the art of geometry." If the applicability of mathematics calls for some explanation, then the mystery is only deepened by taking mathematics to be about a realm disjoint from the natural world.

Faced with these initial hurdles for platonism, one might be tempted to rethink one's strategy for safeguarding empiricism from the *a priori* nature of mathematical knowledge. And there is indeed another measure to hand, namely the proposal that, contrary to all appearances, mathematical truths are in fact known *a posteriori*. Although some philosophers have attempted to elaborate such a view, it is hard to see how they can overcome the fact already emphasized: that mathematical arguments do not seem to be justified by any empirical observations. In addition, the

---

[5] From a lecture delivered to the Prussian Academy of Sciences in Berlin on January 27, 1921: see Einstein (1983, 28).

[6] Weinberg (1986).

very nature of justification in mathematics, namely proof, appears quite different from that usually operative in arguments to conclusions about the natural world. The latter inferences are often inductive or statistical in character, and so not truth-preserving: that is, the truth of their premises does not guarantee the truth of their conclusions, but may make them merely probable to some degree or other. By contrast, the inferences found in mathematical proofs are all *valid*: the truth of such an inference's premises necessitates the truth of its conclusion. It seems, then, that mathematical truths not only fail to be known on the basis of empirical evidence, but they also fail to be justified by inferences of the kind typically used to reason about the natural world. This proposal really does appear a counsel of despair.

This last point draws our attention to yet another feature of mathematics that has puzzled philosophers: the validity of the inferences that figure in its deductions. Consider the form of inference at work in Theorem 1.1. We can analyze it as having the following structure:[7]

(1) $A \rightarrow \neg A$.

Therefore,

(2) $\neg A$.

This argument, known as *reductio ad absurdum*, has the distinctive property that if (1) is true, then (2) must be true as well.[8] The claim is not merely that (2)'s truth is made very likely by the truth of (1) but, rather, that it is made necessary. Although the proof of Theorem 1.2 is supported by a different logical skeleton, it shares this property. Its structure is as follows:

(3) $A \lor \neg A$,

(4) $A \rightarrow B$,

---

[7]   In this book, we use the following logical notation: "$\neg P$" means *not P*, "$P \land Q$" means *P and Q*, "$P \lor Q$" means *P or Q*, "$P \rightarrow Q$" means *if P then Q*, "$P \leftrightarrow Q$" means *P if and only if Q*, "$\forall x P$" means *for all x, P*, and "$\exists x P$" means *there exists at least one x such that P*.

[8]   Hardy memorably said of this form of inference that it "is one of a mathematician's finest weapons. It is a far finer gambit than any chess gambit: a chess player may offer the sacrifice of a pawn or even a piece, but a mathematician offers *the game*." (Hardy, 1967, 94).

(5) $\neg A \rightarrow B.$

Therefore,

(6) $B.$

The unique force of the "therefore" here is just this: that should (3)–(5) be true, then necessarily (6) is as well. The Austrian philosopher Ludwig Wittgenstein (1889–1951) called this the "hardness of the logical *must*,"[9] and many have found its source quite mysterious. For instance, the Scottish philosopher David Hume (1711–76) argued that such a necessary connection between states of affairs is illusory, for nothing like it is given to us in experience; at best, we observe that one kind of state of affairs regularly follows another, but the *necessity* of this link is nowhere to be seen.[10] This necessity is one source of the aesthetic fascination that mathematics exerts. The English philosopher Bertrand Russell (1872–1970) once told a story that illustrates this:

> My friend G. H. Hardy, who was Professor of pure mathematics, enjoyed this pleasure in a very high degree. He told me once that if he could find a proof that I was going to die in five minutes he would of course be sorry to lose me, but this sorrow would be quite outweighed by pleasure in the proof. I entirely sympathized with him and was not at all offended.[11]

The bewitchingly mysterious necessity of proof highlights yet another aspect of mathematics that is difficult to accommodate within an empiricist perspective.

There is a final feature of mathematics, already obliquely mentioned, that deserves comment here. This is that the knowledge justified by mathematical proof is often *infinitary* in nature: the content of what is known involves reference to an infinite range of objects. Consider, for example, Theorem 1.1, whose infinitary content might not at first be obvious, since it seems to be about just one entity, the square root of 2. The fact is that the property being attributed to this entity, that of irrationality, when rendered fully explicit, makes reference to an infinite collection. For to say that $\sqrt{2}$ is irrational is just to say that

$$\neg \exists x \exists y (x \text{ is an integer} \wedge y \text{ is an integer} \wedge \sqrt{2} = x/y).$$

---

[9] Wittgenstein (1978).
[10] See, for instance, Hume (1748).
[11] Russell (1956, 14).

And this is logically equivalent to

$$\forall x \forall y ((x \text{ is an integer} \wedge y \text{ is an integer}) \rightarrow \sqrt{2} \neq x/y).$$

As the universal quantifiers ranging over the totality of integers reveal, the claim that $\sqrt{2}$ is irrational is an infinitary one. Now this feature of mathematics can appear puzzling. We are, after all, finite creatures: our powers, such as our memory and computational speed, are finite, as are the durations for which we can exercise these powers, namely our lives. Consequently, we have the capacity to survey no more than a finite amount of evidence. How, then, do we manage to arrive at infinitary knowledge? One might even begin to find it mysterious that we finite beings are able so much as to *comprehend* statements about an infinite range of objects. As the French writer Voltaire (1694–1778) said, the infinite "astonishes our dimension of brains, which is only about six inches long, five broad, and six in depth, in the largest heads."[12]

Mathematics, in short, seems to be a discipline through which we acquire knowledge about infinitely many entities with which we can in no way causally interact, by means of finite inferences which make no use of empirical premises and which yield their conclusions with the force of necessity. We began by noting that it might seem surprising that mathematics should be the object of so much philosophical attention. But perhaps what should surprise is any display of intellectual equanimity before the phenomenon of mathematics.

Philosophers have not been alone in trying to understand mathematics, however distinctive their own perplexities and approaches might be. Mathematicians, spurred especially by dissatisfaction with their understanding of analysis, the mathematics that underpins the calculus, have in their own way sought to arrive at a deeper understanding of the foundations of their discipline. These attempts were especially sustained in the nineteenth century, although the worries dated from before. The Irish philosopher Bishop Berkeley (1685–1753), complaining in his *The Analyst, or a Discourse Addressed to an Infidel Mathematician* (1734) about Newton's formulation of the calculus, quipped that: "He who can digest a second or third fluxion, a second or third difference, need not, methinks, be squeamish about any point in divinity."[13] Although many struggled to clarify these matters, the mathematician N. H. Abel (1802–29) could still complain in 1826, almost a century later, of "the surprising obscurity one finds undoubtedly in analysis today. It lacks all plan and

---

[12]   Quoted in Moritz (1958, 336).
[13]   Reprinted in Ewald (1996, vol. 1, 62–92; 65, para. 7).

unity [...] the worst is that it has not at all been treated with rigor. There are only a very few theorems in advanced analysis which are proved with complete rigor."[14] Indeed, at that time mathematicians were still lacking definitions of basic analytic notions such as limit, continuity, the distinction between pointwise and uniform convergence, and the derivative.

Yet by the end of the nineteenth century, conceptual clarity regarding the foundations of analysis had been largely achieved. The rigor whose absence was lamented by Abel was now in place, and it made possible not only a deeper understanding of previous results (and errors), but also suggested new, formerly unthinkable avenues of exploration. This enhanced focus developed as mathematics, the calculus especially, underwent a systematic analysis in terms of the familiar natural numbers and the usual operations defined on them. Of particular note in this connection is the accomplishment, due primarily to the German mathematician Richard Dedekind (1831–1916), of defining the integers, rationals and reals, taking only the system of natural numbers for granted.

It suffices now to say that for some this process of analysis looked to be at an end, on the grounds that there was nothing more basic than the natural numbers to which they in turn could be reduced. In this spirit, the German mathematician Leopold Kronecker (1823–91) famously announced that "God made the integers, all the rest is the work of man."[15] This attitude does not entail that no further progress can be made in understanding the basis of mathematics, but only that we should not look to mathematics itself for deeper illumination; in the natural numbers, we have hit mathematical rock-bottom. Should still greater insight be achievable through a philosophical analysis of natural number, someone with this attitude might urge, we must not expect it also to provide anything that could pass for a mathematical account.

In the next chapter, we shall consider how in fact some sought to achieve both goals simultaneously.

[14] Abel (1902, 23), quoted in Sieg (1984).
[15] Quoted in Bell (1937, 477).

2

Logicism

Gottlob Frege (1848–1925), a German mathematician who was trained in physics and philosophy, and who spent his entire working life at the University of Jena, was the first to describe, develop, and defend in detail what has since come to be known as a logicist approach to mathematics. Before saying what this consists of, we would do well to relate briefly what Frege found unacceptable about many of the then prevalent views about the nature of mathematics.

By Frege's time, much had been clarified regarding the relationships among the different number systems. In particular, it was understood how all the notions pertaining to the systems of the real numbers, the rationals, and the integers could be defined by reference to properties of the natural numbers (the numbers 0, 1, 2, and so on), and also how truths regarding the former kinds of number could be established on the basis of principles true of the natural numbers. Precisely how this *arithmeticization* of analysis proceeds is something to which we shall turn in the next chapter.

Unsurprisingly then, Frege first directed his attention to the status of the natural numbers, and his *The Foundations of Arithmetic* (1884) opens with the question: What is the number one?[1] In that work, Frege makes it plain that he thinks that prevalent views about the nature of natural numbers and of arithmetical justification founder through their focus on experience, whether on its content or its form. By the content of experience, we mean the particular sensations that humans have through the exercise of their sensory organs. Although, as we noted in the previous chapter, some – like the English philosopher John Stuart Mill (1806–73) – have been tempted to claim that mathematical truths in general are ultimately justified on the basis of our experience,[2] Frege

---

[1] Frege (1884, 10e).
[2] Mill (1843).

rejects the view that the contents of our experience are at all relevant to the truth of arithmetical claims. By the form of experience, we mean the structure to which all human experience must conform, for example, as of events or objects located in time or organized within a particular kind of space. Such structural constraints underpin many of our intuitions, for example the spatial intuitions that we often draw on when solving certain kinds of geometric problems. While some, like the German philosopher Immanuel Kant (1724–1804), have held that arithmetic is in some manner based on such intuitions,[3] Frege is adamant that they have no role to play in so far as justification is concerned. That our experience is structured in a particular way might be relevant to understanding how we come to discover certain truths, or why we find some truths easier to grasp than others, but it cannot provide us with rational grounds for arithmetical truth.

But if sensation and intuition cannot, what can? No answer lay ready-to-hand for Frege, and so he had to invent one. In 1879, Frege published his *Begriffsschrift*, or *Concept-script*, a thin volume in which quantificational predicate logic made its first appearance.[4] Logic, for Frege, offers a precise systematization of the canons of reason. It takes as its subject matter all valid inferences, regardless of the particular content of the statements they involve.[5] By using the quantifier–variable notation and by replacing the grammatical subject–verb analysis by a more functional one, Frege's logic makes perspicuous the hidden structure of statements in virtue of which they make the claims they do and are inferentially related to one another in the ways they are. Logic, according to Frege, limns the laws of thought. These must not be confused with the laws of thinking, the psychological processes that might take place as one reasons. Rather, they are the laws that constrain what can be rationally thought. Logic does not aim to describe how humans think, but instead to characterize how they must think if their thought is to remain within the bounds of reason.[6]

Furthermore, this characterization is to be effected in a completely systematic fashion. In showing how this could be achieved, Frege made the greatest contribution to the development of the axiomatic system since the Ancient Greeks. He did this by carefully isolating and describing

---

[3] See, for example, Kant (1783).

[4] See van Heijenoort (1967, 1–82).

[5] For instance, see his "Logic" (1882–8), in Hermes, Kambartel, and Kaulbach (1979, 3).

[6] For instance, see his "Thoughts" (1918), in Salmon and Soames (1988, 33–55).

his system's basic premises and rules of inference. Although he does not view his logic as an uninterpreted calculus, it is important for him that it be a *formal system* in that its licensing of a putative demonstration be amenable to a thoroughly formal check. Proofs of even the most basic of assertions can be lengthy (as students of logic will have noted), but this is compensated for by the fact that the number of axiom forms and rules of inference is small. Although perhaps long, all proofs are recognizably free from gaps and leave no room for doubt as to whether they are proofs.[7] Actually, to say that this is a compensation for the length of proofs in Frege's logical system is to misidentify the latter's role: it is not meant as a tool for the working mathematician in his or her everyday research, but rather as a contribution to our understanding of inference, and to our ability to clarify and order any given domain of knowledge by fully revealing the systematic justificatory structure that relates its claims.[8]

With logic in hand, there is now a way to settle the issue of the foundational basis of arithmetic. Using the principles of logic, one sets out to organize all arithmetical knowledge. Specifically, one tracks the basis of knowledge backwards until one arrives at a collection of assumptions or principles of reasoning that are not themselves subject to justification. This collection would constitute the justificatory source of arithmetic: from it, together with logic, all arithmetical knowledge would be derivable. The fundamental nature of arithmetic could then finally be determined by inspecting the elements of its source.[9] This procedure makes it plain why Frege insists that logic provide a way of constructing gap-free demonstrations: without this assurance, we would have no reason to believe that the ultimate justification of a science depends on just the principles isolated by this procedure, and so no reason to believe that scrutiny of these principles would reveal anything fundamental about the nature of the science.

Frege claims to have carried out this procedure for arithmetic, and he advances a bold and startling thesis regarding the collection of assumptions and principles from which all of arithmetic could be logically derived: it is empty. Put otherwise, Frege argues that nothing is needed to justify our arithmetical knowledge beyond logic's axioms and its rules of inference. There are no distinctively mathematical postulates or rules of inference that have to be adopted; those of logic, those already

---

[7]   See his "Boole's logical calculus and the concept-script" (1881), in Hermes et al. (1979), pp. 32, 39.

[8]   See his "Logic in mathematics" (1914), in Hermes et al. (1979), p. 242.

[9]   For example, ibid., p. 205.

explicitly articulated in the logical system, suffice. Arithmetic, Frege argues, is ultimately based on logic: all truths of arithmetic can be analyzed using purely logical notions and proved from logical axioms on the basis of logical principles of inference. Despite its surface appearance, arithmetic is actually a direct development of the science of inference; it depends for its truth only on the laws of rational thought taken most generally. This thesis has come to be known as Frege's *logicism*.

Logic thus provides Frege with the sought-after counterpoint to the foundational claims of experience. Although he offers many reasons for thinking that experience, either its form or its content, is not implicated in the foundations of arithmetic,[10] behind these negative considerations stands a central positive reason for rejecting the foundational relevance of experience: the realizability of the logicist project. The latter directly shows, Frege claims, that no empirical truths about the natural world need be employed in the justification of arithmetic, nor need any truths that are apprehended through some kind of intuition. His negative arguments highlight the unsatisfactory nature of experience-based accounts of arithmetic, while the logicist reduction reveals that all such proposals are unnecessary in the first place.

It is worth underlining here the central role that logic plays in Frege's approach. Most obviously, of course, it is the answer to the question he is intent on posing: "What are the foundations of arithmetic; that is, which truths do the truths of arithmetic depend on for their justification?" But it bears noting that logic also shapes the very content of his question. For Frege's understanding of what it is for something to be the justificatory source of arithmetic is thoroughly conditioned by his conception of logic: nothing will count as such a source unless all of arithmetic can be deduced from it through derivations sanctioned by logic. There is an important sense, then, in which, prior to Frege's logical discoveries, not only could his central question not have been answered, but it could not even have been posed.

We just saw that a successful execution of the logicist program would lead to greater clarity regarding the justificatory basis of arithmetic. Frege particularly stresses that the reduction would demonstrate that arithmetic's justification requires no reference to psychology. This assumes of course that logic, what arithmetic can allegedly be reduced to without residue, does not in turn depend on any psychological principles. Frege insists that this is the case: logic is the study of truth-preserving inference, and to think that it might have to appeal to psychology is

---

[10]   See especially Part I of his *Foundations*.

simply to misunderstand the word "'true', which excludes any reference to a knowing subject."[11]

Frege's response brings out clearly that the philosophical illumination a successful logicist reduction can be expected to shed will depend on how one views logic itself: the reduction of arithmetic to logic in effect transforms the questions canvassed in the previous chapter into questions about logic. If logic were merely another branch of inquiry, one would think that most of these could be raised with no less force with respect to it. But Frege does not conceive of logic in this way. Although logic is to be pursued with the same systematicity and rigor as other sciences are, it does not sit at the same level of generality as they do. It does not, like other disciplines, study a specific range of entities or phenomena, and seek after truths concerning them. Rather, its focus is truth in general: it strives not to define the predicate "true" (something Frege believes to be impossible), but to characterize the conditions under which truth is transferred from proposition to proposition. The laws of logic govern all rational thought, regardless of its particular content: logic "[disregards] the particular characteristics of objects, [and] depends solely on those laws upon which all knowledge rests."[12] The laws of logic are what make it possible to infer one sentence from another, to justify one claim by appeal to another. Consequently, there would be something misplaced in a request for a wholesale justification of logic itself. To engage in reasoning is to rely on certain forms of argument, which already requires acceptance of a system of logic. To ask for a justification of logic is, as Frege puts it, to want "to judge without judging, or to wash the fur without wetting it."[13] For this reason, a logicist reduction, understood as Frege intends, would lay to rest general questions about the justification of mathematics.

There are other respects in which logism alleviates some of the mysteries about mathematics scouted in the previous chapter. We saw there that one might easily wonder how arithmetical knowledge could be acquired solely on the basis of reflection. How can one, merely by thinking, acquire information about the natural numbers? But if logicism is correct, then an arithmetical truth is ultimately a logical truth, a truth about the constraints on rational thought. And there is perhaps less mystery in seeing how such truths might be accessible to pure thought: they concern the laws that govern its own functioning, instead of those

---

[11]   "Logic," p. 5. See also Frege's "Logic" (1897), in Hermes et al. (1979), pp. 132–3.

[12]   *Begriffsschrift*, p. 5.

[13]   *Foundations*, p. 36e.

that regulate some domain quite disconnected from it. "In arithmetic," Frege says, "we are not concerned with objects which we come to know as something alien from without through the medium of the senses, but with objects given directly to our reason and, as its nearest kin, utterly transparent to it."[14] Earlier, we also raised the puzzling applicability of mathematics. But if mathematics is logic, and logic is the most general of all disciplines, one that applies to all rational thought regardless of its content, then it is not surprising that mathematics is so widely applicable.

Frege holds that a successful reduction would not only be philosophically valuable, but mathematically illuminating as well. The mathematical value would be primarily two-fold. In the first place, such a reduction would reveal with complete clarity the relations of logical dependency among the truths of arithmetic. It may happen, for example, that we learn that a theorem thought to depend on a collection of hypotheses in fact depends on some proper subset: one or more hypotheses is thereby shown to play no role in guaranteeing the theorem's truth. Or it might be discovered that a collection of hypotheses that was thought to be sufficient for the truth of a claim in fact is not. These are the kinds of situations that will be exposed by a reduction using Frege's system of logic, and their discovery must count as mathematical progress.

There is a second respect in which Frege thinks that a reduction would be of mathematical value. For not only would it enhance clarity regarding the logical relations of claims to one another, but it would also lead to a greater understanding of the mathematical concepts involved in those claims. As ever, Frege is not interested in matters of psychology here: just as his concern is not with how we happen to know or discover arithmetical truths, but rather with the relations of justification that hold among them, so his interest is not in the psychological, biological, or physical preconditions for actually arriving at concepts. His focus is rather on the concepts themselves and on their organization, which they have independently of us, independently of our beliefs. The concepts of arithmetic, in particular, exhibit an extreme "richness and fineness in their internal structure,"[15] the articulation of which can be difficult and yet necessary for mathematical progress. "Often," Frege points out, "it is only after immense intellectual effort, which may have continued over centuries, that humanity at last succeeds in achieving knowledge of a concept in its pure form, in stripping off the irrelevant accretions which veil it from the eyes of the mind."[16] The result

[14] *Foundations*, p. 115.
[15] Hermes et al. (1979), p. 13. See also *Foundations*, p. iv.
[16] *Foundations*, p. vii.

of such effort can be an extension of mathematical knowledge, and a deepened understanding of what is known.

Now just as Frege's notion of justificatory dependence between truths is intimately bound up with his articulation of logic, so too is his notion of a concept. Concepts, for Frege, are the ontological counterparts of predicative expressions. We shall use the term *names* to refer to all expressions like proper names (for example, "Frege") and definite descriptions (for example, "the man who wrote *Begriffsschrift*"). Let us, with Frege, call what names refer to *objects*. The result of removing a name from a declarative sentence is a predicative expression that refers to a *concept*; concepts just are what predicative expressions denote. For example, if we remove the name "Kant" from:

(1)   Kant was born before Frege,

we obtain the predicate "__ was born before Frege." This predicate is true of some objects (such as Plato) and false of others (such as Frege). We shall say that the objects of which the predicate is true *fall under* the concept to which this predicate refers. We could have removed a different name from (1) to arrive at a different predicate; for example, we could have instead formed the predicate "Kant was born before __." This is another *one-place* predicate, because it is formed by the removal of one name. Concepts referred to by such predicates can be viewed as properties of objects; for instance, the concept corresponding to the predicate "__ was born before Frege" is simply the property of having been born before Frege. We are free, however, to remove any number of names, thereby forming predicates with different numbers of argument places; for example, from (1) we can form the *two-place* predicate "__ was born before __." This likewise refers to a concept, this time to one under which fall pairs of objects (such as ⟨Plato, Wittgenstein⟩). Such two-place concepts are also called *relations*.

Before we turn to Frege's analysis of arithmetic, three further observations about concepts are in order. First, because a concept is just what is denoted by an expression that has been formed by removing a name from a sentence, concepts are different from objects in being inherently incomplete in nature. Unlike objects, they contain a gap that requires supplementation, much as a function always carries an argument place that must be filled before it yields a value. Indeed, Frege views concepts quite simply as functions from objects (or pairs of objects, and so on) to truth values – of which there are two: the True and the False.[17] For instance,

---

[17]   See his "Function and concept" (1891), in Geach and Black (1980, 28).

the concept denoted by "__ is German," when supplied with Frege as argument, yields the value True, but when supplied with Hume yields the value False. In mathematics, functions are sometimes identified with what Frege would consider objects. For example, the real-valued function $f(x) = x^2$ might for some purposes be taken to be the set $\{\langle x, x^2 \rangle : x$ is a real number$\}$. (We will have more to say about this identification in the next chapter.) Frege would insist, however, that we sharply distinguish between a concept F and the collection of objects that fall under F. The latter collection he calls the concept's *extension*; it is an object, as is revealed by the fact that we use a name ("the concept F's extension") to refer to it.[18] Although F and its extension are related, they cannot be identified. Likewise, even if it should turn out that but a single object falls under some concept, we must still distinguish among the concept, the object that falls under it, and finally the extension of that concept, which is the set containing that one object. For example, as it happens there is only one object that falls under the concept denoted by "__ is a heavenly body that orbits the Earth," yet the two must be kept distinct: the concept is an incomplete entity, whereas the Moon is a complete object. And each of these must in turn be distinguished from the extension of the concept, that is, from the set whose sole member is the Moon: this set is not identical to its only element and, being an object, it is also not identical to the concept of which it is the extension.

Secondly, we can now offer a few words to clarify our earlier remark that Frege's notion of concept is inextricably related to his understanding of logic. Frege insists that the appropriate analysis of sentences is in terms of name and predicate, because this analysis is what is needed for an account of the inferential connections between sentences. Frege's emphasis on the importance of the notions object and concept follows from this, since they are the ontological correlates of these types of expression. It is one of Frege's great contributions to recognize that only by viewing a sentence as decomposable into predicates can we account for the inferences in which it takes part. For instance, in order to see why (1) entails

---

[18] The reader might well wonder whether, by this criterion of concept-hood, Frege is committed to claiming that "the concept denoted by '__ is German'" does not in fact refer to a concept. Indeed, this is precisely the conclusion he draws. Contrary to appearances perhaps, the entity it denotes does not contain any gaps in the way the entity denoted by "__is German" does; the first entity, unlike the second, cannot take anything as argument. For further discussion, see his "On concept and object" (1892), in Geach and Black (1980, 42–55).

(2)   Someone was born before Frege,

or why it is entailed by

(3)   Kant was born before everyone,

we need to recognize that both (1) and (2) can be viewed as constructed from the predicate "__ was born before Frege" (so as to see that the first inference is an instance of existential generalization), and that both (1) and (3) can be analyzed in terms of the predicate "Kant was born before __" (to see that the second inference is a case of universal instantiation). The decomposition of sentences into predicates that refer to concepts is bound up with the goal of arriving at an analysis of inferential dependence.[19] If we keep this notion of concept in view, we can see why Frege insists that a logical reduction of arithmetic could not but illuminate the concepts involved in arithmetical truths.

Finally, we observe that the only concepts considered so far are those under which objects can fall; we will call these *first-level concepts*. We saw that a first-level concept is referred to by an expression obtained by removing a name from a declarative sentence. We will call such an expression a *first-level predicate*: it is one that contains a gap that can be filled by a name. First-order quantificational logic restricts itself to first-level predicates; it countenances quantification over objects only. But, starting with a complete sentence, we can also remove now not a name but a first-level predicate. The resulting incomplete expression is a *second-level predicate*; it denotes a *second-level concept*, which is one under which first-level concepts can fall. We might view second-level concepts as properties of properties of objects. Predicates and concepts of ever higher levels can be defined as well. For Frege, logic is not first-order: it admits predicates of higher level and quantification over concepts of higher level as well. In Frege's logic, one can quantify not only over objects, but over first-level concepts, second-level concepts, and so on. Consider, for example,

(4)   All dogs are mammals.

The logical structure of this sentence is as follows:

(5) $\forall x(x$ is a dog $\rightarrow x$ is a mammal).

---

[19]   See, for example, his *Begriffsschrift*, §9; also Hermes et al. (1979), pp. 16–17.

If we remove the first-level predicate "__ is a dog" we arrive at

(6) $\forall x(\ldots x \ldots \rightarrow x$ is a mammal).

This is a second-level predicate. Depending on the predicate of first level that we use to fill its gap, a true or a false sentence results. For instance, if we insert the first-level predicate "__ is a dolphin" a truth results, namely "$\forall x(x$ is a dolphin $\rightarrow x$ is a mammal)"; if we insert "Frege wrote __" a falsehood does, namely "$\forall x$ (Frege wrote $x \rightarrow x$ is a mammal)."[20] Thus (6) is a predicate that denotes a concept of second level, that is, one under which first-level concepts fall.

On this analysis, the assertion of existence is likewise one that involves a second-level concept. We arrive at the existential quantifier by starting with an existence claim – for instance, "Something is a black hole" – and removing a first-level predicate, in this case "__ is a black hole." The existential quantifier, "there is at least one x such that $\ldots x \ldots$," can now be seen to be a second-level predicate; it is true of just those first-level concepts that are true of at least one object (for example, the concept denoted by "__ is a philosopher"). The contribution of the quantifier to the truth conditions of sentences of which it is a part cannot be adequately explained if it is treated as other than a second-level predicate, for instance, if it is viewed as a name. To do so would thwart an analysis of existential statements that explains why they function as they do in logical inferences.

We are now in a position to examine Frege's analysis of the natural numbers. Frege notes that number words ("one", "two", and so on) often appear as adjectives, as when we say that there are ten Plagues of Egypt. Such assertions seem to attribute a property – but to what? The superficial similarity between "The Plagues of Egypt are deadly" – which does impute the property of being deadly to each of the Plagues of Egypt – and "The Plagues of Egypt are ten" suggests that it might be attributed to objects, to the plagues themselves. However, this cannot be correct: after all, the plague of locusts may be deadly, but it makes no sense to say that that plague itself is ten.

Frege argues that a numerical adjective, such as appears in "There are ten Plagues of Egypt," forms a part of a predicate of second-level, that is, one that requires supplementation by a first-level predicate, in this case

---

[20] Note that after we remove "__ is a dog" from (5), a bound variable remains behind, for the variable is not part of the predicate. When we insert a first-level predicate into (6), this variable fills its gap, wherever the gap is located.

"__ is a Plague of Egypt."[21] This second-level predicate is therefore of the same type as the existential quantifier, and we signal this by calling it a numerical quantifier, which we can also write "There are exactly 10 x's such that...x...." The concept of being ten in number is thus actually a second-level concept, one under which first-level concepts may fall (such as *Biblical Commandment*, but not *U.S. President*).[22] Frege was the first to reveal the logical structure of the second-level concepts that correspond to the adjectival use of number. His characterization is *recursive*; that is, it is given in stages, each one defined in terms of the previous one:[23]

(7a) $\exists_0 xFx$ ["There are no Fs"] means $\neg\exists xFx$,

(7b) $\exists_{n+1} xFx$ ["There are exactly $n + 1$ Fs"] means
$\exists x(Fx \wedge \exists_n y(Fy \wedge x \neq y))$.

For example, according to this analysis,

(8)   There are exactly two kings

has the logical structure

(9a) $\exists_2 x(x$ is a king).

Applying (7b) with $n = 1$, we have that this in turn means

(9b) $\exists x(x$ is a king $\wedge \exists_1 y(y$ is a king $\wedge x \neq y))$.

Another application of (7b), this time with $n = 0$, yields

(9c) $\exists x(x$ is a king $\wedge \exists y(y$ is a king $\wedge x \neq y \wedge$
$\exists_0 z(z$ is a king $\wedge x \neq z \wedge y \neq z)))$.

Finally, we can apply clause (7a) to yield the full logical analysis of (8):

(9d) $\exists x(x$ is a king $\wedge \exists y(y$ is a king $\wedge x \neq y \wedge$
$\neg\exists z(z$ is a king $\wedge x \neq z \wedge y \neq z)))$.

---

[21]   For Frege's discussion, see *Foundations*, §§51–3.
[22]   As a convention, we shall sometimes use "*Biblical Commandment*" to refer to the concept denoted by "__ is a Biblical Commandment"; and similarly for other instances.
[23]   *Foundations*, §55.

This is logically equivalent to

(10) $\exists x \exists y (x$ is a king $\land y$ is a king $\land x \neq y \land$
$\forall z (z$ is a king $\rightarrow (z = x \lor z = y)))$,

which is perhaps slightly more perspicuous. If we remove all occurrences of the predicate "__ is a king" from (10), we get a numerical quantifier: a predicative expression that refers to a concept of second level. This concept, *two in number,* is one under which fall some first-level concepts (for example, *side in chess*) but not others (for example, *inhabitant of Amherst*). It can be viewed as a property, not of objects but rather of properties of objects: it is the property possessed by all and only those properties that exactly two objects have. Frege's recursive definition (7) allows us to articulate within the language of logic, for each *n*, the complete structure of the numerical quantifier "There are exactly *n* x's such that ...x...."

Furthermore, not only does Frege's analysis permit one to see the logical structure of assertions about number, it also permits one to derive arithmetical truths using just the resources of logic. For instance, the reader might like to check that the following statement is derivable by the standard laws of logic:

$$[\exists_5 x Fx \land \exists_7 x Gx \land \neg \exists x (Fx \land Gx)] \rightarrow \exists_{12} x (Fx \lor Gx).$$

Because this statement intuitively expresses that $5 + 7 = 12$, we see that this identity is in fact a logical truth.

Does this analysis of numerical quantifiers count then as a logicist reduction of arithmetic? Frege argues that it does not, because by itself it fails to provide the means to express the content of such assertions as

(11) The number of Plagues of Egypt is equal to the number of Biblical Commandments.

It is quite true that both concepts *Plague of Egypt* and *Biblical Commandment* fall under the second-level concept denoted by the numerical quantifier "There are exactly 10 x's such that...x....," and consequently it is true that there is some second-level concept under which they both fall. (While this cannot be expressed in first-order logic, it can be in Frege's logical system which, as earlier noted, allows quantification over concepts of higher level.) But that consequence falls short of the content of (11): after all, any two first-level concepts under which fall some objects – although not necessarily the same number of objects – will both fall under some second-level concept: that of exist-ence, for example.

We can of course say that there are exactly ten Plagues of Egypt and exactly ten Biblical Commandments:

(12) $\exists_{10}x(x$ is a Plague of Egypt$) \land \exists_{10}x(x$ is a Biblical Commandment$)$.

But this does not explicitly capture the identity of number expressed in (11). We might say that (12) *shows* that the identity holds but fails to *say* that it does.

The identity in (11) can also be expressed as the claim that there exists a number which is both the number of Plagues of Egypt and the number of Biblical Commandments:

(13) $\exists n(n = $ the number of Plagues of Egypt $\land \ n = $ the number of Biblical Commandments$)$.

There is now a great temptation to read "$\exists_{10}x$ (x is a Plague of Egypt)" as "$10 = $ the number of Plagues of Egypt." If we allow ourselves this reading, then we can express (13), and hence our original (11), by existentially quantifying (12):

(14) $\exists n(\exists_n x(x$ is a Plague of Egypt$) \land \exists_n x(x$ is a Biblical Commandment$))$.

The problem is that we have no way of understanding this statement using the analysis of the numerical quantifiers in (7). Once "$\exists_{10}x(x$ is a Plague of Egypt)" is replaced by its definition, the numeral "10" disap-pears, and so there is no position for the outermost quantifier in (14) to quantify into. The temptation is in fact one to which we should not succumb: the notation employed in (7) suggests that we are defining a complex expression "$\exists_{10}$", one of whose components is the numeral "10". In fact, once we examine the logical analysis, we see that that is not so, and that the subscripted quantifier must be treated as a logically simple symbol. What (7) succeeds in defining is "$\exists_0 x$", "$\exists_1 x$", "$\exists_2 x$", and so on; but it does not define "$\exists_n x$", where "n" is a variable.

Frege says that (7) will not permit us to express (11) because the latter makes reference to numbers viewed as objects, whereas the former, despite tempting appearances to the contrary, does not. For Frege, (7) is not an adequate analysis of number because it fails to provide an account of natural numbers as objects. In daily life, but especially in mathematics, numbers are typically referred to using names, either descriptively (for example, "the number of clubs in a deck") or by employing canonical proper names (for example, "13"). The mathematician usually thinks and speaks of numbers as if they were individual entities, each with its own distinct cluster of properties. Frege says this provides a reason to treat numbers as "self-subsistent objects,"[24] and he holds that any adequate reduction of arithmetic to logic would have to say exactly what these objects are.[25]

For this reason, Frege suggests that before we offer any analyses we rephrase "There are exactly 10 Plagues of Egypt" so as to bring out the objectual nature of number. We begin by noting that the meaning of this statement is the same as that of "The number of Plagues of Egypt is 10," where the "is" signals not predication but an identity claim. The form of "The number of Fs" is a name, and so of the right kind to denote an object. Spelled out, its structure is: the number of x's such that Fx. We can view "the number of x's such that $\dots x \dots$" as a functional expression that is completed by a first-level predicate and yields a name. The task now is to offer a logical analysis of the meaning of this structurally complex term. But how is such an analysis to proceed?

In asking this question, Frege took a step that stamped itself on much of twentieth-century philosophy: he turned a question about the nature of particular entities – in this case, the natural numbers, viewed as objects – into one about the meaning of particular expressions – in this case, names of the form "the number of Fs." In answering this question, Frege made a suggestion that likewise came to dominate subsequent reflection on language. For Frege proposed that we should *not* analyze the meaning of these expressions on their own. Any inquiry into the meaning of an expression in isolation is fraught with the danger of identifying its meaning with the mental images it might conjure up in us; thus many people,

---

[24] *Foundations*, §56.

[25] Both conjuncts of (12) contain as a constituent the second-level numerical quantifier "There are exactly 10 x's such that $\dots x \dots$." It is therefore open to Frege to quantify into that predicate position using a third-level expression. Assume that "M—" is a third-level predicate that is true of all and only second-level concepts denoted by a numerical quantifier. Then Frege could express (12) by saying that there exists a second-level concept of which "M—" is true and under which both *Plague of Egypt* and *Biblical Commandment* fall. This approach, however, fails to provide Frege with an analysis of numbers as *objects*, something he insists on. Below, we will explore a more structural reason why it was important to Frege that numbers be objects. For a very useful discussion of this, as well as many other aspects of Frege's work on mathematics, see Dummett (1991, in particular 131–4).

when asked about the meaning of, say, the word "dog," will talk about the pictures that come before their minds. Such an identification risks leading into the kind of psychological speculation that Frege insists is thoroughly irrelevant here. More deeply, Frege holds that this procedure of analysis misunderstands what the meaning of an expression is. According to him, the fundamental unit of meaning is that of the whole sentence. It is only through the use of a complete sentence that one can express a thought. For this reason, he insists that there is nothing more to be said about the meaning of an expression beyond a characterization of how it contributes to the meanings of sentences in which it appears. Note that it could be misleading to characterize Frege's procedure as an indirect inquiry into an expression's meaning. For this might encourage the thought that there is something, the expression's meaning, which for some reason is not amenable to direct analysis. But for Frege, an account of how an expression contributes to the meaning of full sentences in which it occurs *is* a direct and exhaustive account of the expression's meaning. Through this emphasis on *the primacy of the sentence*, Frege set the course for much of the twentieth century's philosophical reflection on language.

One immediate problem facing such a procedure is that any expression will be part of infinitely many sentences of all different kinds of structure. One cannot check each of these to see how the expression in question contributes to the sentence's meaning. Frege suggests that if the expression whose meaning we are seeking to characterize is a name, then there is a particular sub-class of sentences in which it appears to which one must pay special attention. These sentences Frege calls "recognition-statements";[26] they are identity statements in which one of the expressions flanking the identity symbol is the kind of name whose meaning one is trying to characterize.[27] Thus, in the case at issue, recognition-statements will have the form

(15)   The number of Fs $= k$,

where "k" is any name. If we are to succeed in saying what the meaning is of expressions of the form "the number of Fs," then we must give an analysis of the meaning of (15) that shows how the occurrence of expressions of that form contributes to the identity's content.

---

[26]   *Foundations*, §§62, 106.

[27]   Of course, there will be infinitely many recognition-statements in which a given name appears. But the hope is that one will be able to say in general how that term contributes to the meaning of all such statements.

Before proceeding, we should say a word about why Frege thinks it necessary to give an analysis of the meaning of recognition-statements in particular. Frege, we saw, wants an account of natural numbers that treats them as objects. We cannot consider something to be an object, however, unless we can, if only in principle, distinguish it from other things. Now in order to be able so to distinguish it, we must be able to re-identify it. For if one is in principle incapable of determining when one is presented with the object o again, then one will likewise be unable to determine whether one is being presented with an entity other than o – and this inability betrays one's lack of an objectual conception of o. The linguistic correlate of this observation is that in order for an expression to be a genuine name, we must understand the meaning of recognition statements in which that expression occurs.

Frege begins by restricting his attention to recognition-statements of the form

(16) The number of Fs $=$ the number of Gs.

And his analysis of this identity is just that it holds if and only if

(17) There exists a one-to-one correspondence between F and G,

that is, there exists a relation R such that every object that falls under F is related by R to a unique object that falls under G, and vice versa. If (17) holds, let us say that F and G are *equinumerous*. This is a compelling analysis: it is clear that the number of men in a room is identical to the number of women just in case we can pair up the men and the women without remainder. Such a correspondence is a relation between objects, and hence is a relation of first level. The existence of a first level relation with the required properties is something that can be stated using just the resources of higher-order logic, say as follows:

$$\exists R[\forall x(Fx \rightarrow \exists_1 y(Rxy \wedge Gy)) \wedge \forall y(Gy \rightarrow \exists_1 x(Rxy \wedge Fx))].$$

So this analysis remains, for Frege, "a part of pure logic."[28] The analysis does not provide an *explicit definition* of "the number of Fs"; that is, it does not offer a term that can be substituted for "the number of Fs" wherever it might occur (as, for example, "female fox" can be for "vixen"). Rather, the analysis is a partial *contextual definition* of "the number of Fs": it shows how to analyze the expression *in situ*, by replacing a

[28] *Foundations*, p. 83.

complete sentence (of a particular form) in which the expression occurs by another in which it does not.

As we shall see momentarily, this analysis of (16) suffices for the derivation of arithmetic from logic. Frege, however, finds that, as a complete explanation of recognition-statements involving expressions of the form "the number of *Fs*," it is lacking. For (17), he observes, is available only if the two names flanking the identity symbol are of the form "the number of *Fs*"; that is, it is available for recognition-statements of the form (16), not for those of the more general form (15). To take Frege's example, (17) does not tell us how to evaluate the statement "The number of *Fs* = Julius Caesar." Of course, we know that, for any concept *F*, this statement is false; but, as Frege points out, this is no thanks to the definition (17).

This consideration leads Frege after all to offer an explicit definition of "the number of *Fs*":

(18)   The number of *Fs* = the extension of the concept *equinumerous with F*.

The extension of such a concept will be a collection of first-level concepts, namely, just those that are equinumerous with *F*. For example, the number of aces in a deck of cards will be the collection of those concepts that are equinumerous with the concept *ace in a deck of cards*; that is, it will be a collection of concepts that includes the concepts *ace in a deck of cards, Horseman of the Apocalypse, season of the year, rook in a chess set*, and so on, for between each of these concepts and *ace in a deck of cards* there exists a one-to-one correspondence. Frege shows that, given this explicit definition, (16) is logically equivalent to (17); in other words, his partial contextual definition of "the number of *Fs*" follows from his explicit definition (see exercise 2).[29]

It is worth pausing for a moment to remark on Frege's definitional device. The relation of equinumerosity defined by (17) has the interesting property that it effects a partitioning of the universe of concepts into clusters: all the concepts under which exactly one object falls get grouped together into one cluster; all the concepts under which exactly two objects fall get grouped together into another cluster; and so on. Concepts in the same cluster are equinumerous with each other, while concepts from different clusters are not. Since two concepts have the same number assigned to them if and only if they are equinumerous, we can think of Frege's task, of giving an explicit definition of numbers, as

[29]   *Foundations*, §73.

requiring that he associate a single number with each cluster of concepts. His solution is brilliant in its simplicity: he defines the numbers to be the clusters themselves!

Relations that determine such clusters arise often in mathematics. They are called *equivalence relations*, and the clusters they determine are called *equivalence classes*. In the following chapter, we shall treat of equivalence classes more fully, and here we simply note that Frege was one of the first to appreciate their utility, and to use them self-consciously as a definitional tool. (For a preview of the ideas behind equivalence relations, see exercise 1.)

Despite its interest, Frege's explicit definition (18) presents both exegetical and conceptual problems. First, it is unclear why, after Frege has extolled the virtues of contextual definition, he feels the need to offer an explicit definition of "the number of Fs." Furthermore, if he *is* eventually going to offer an explicit definition, why, one might well wonder, does he bother to digress at such great length about a contextual analysis? Perhaps the detour through the analysis of (16) in terms of (17) was thought by Frege to be necessary in order to provide a constraint on any explicit definition; perhaps, that is, the capturing of the equivalence of (16) and (17) functioned for Frege as an adequacy condition on any explicit analysis of "the number of Fs." This might explain why, given that Frege intends to provide an explicit definition, he still goes out of his way to offer a partial contextual analysis of "the number of Fs."

At any rate, it redirects our attention to the first exegetical question, to why Frege finds it necessary to offer an explicit analysis at all. The question is sharpened by the fact that after the explicit definition is employed to prove the equivalence of (16) and (17), it is no longer used: the rest of the reduction takes place solely on the basis of the equivalence. What Frege says, we saw a moment ago, is that an explicit definition is needed so that we know how to evaluate recognition-statements only one of whose names is of the form "the number of Fs"; on the basis of the equivalence of (16) and (17), we could not evaluate the truth value of, say, "The number of cards in a deck is equal to Julius Caesar." And our inability to do this compromises our claim to have provided an objectual analysis of the natural numbers.

In taking his explicit definition to solve these problems, Frege must be assuming that, by contrast, a recognition-statement only one of whose terms is of the form "the extension of G" is one that we *are* able to evaluate. If Frege had not thought that we could evaluate the truth value of, say, "The extension of the concept denoted by 'equinumerous with the concept *card in a deck*' is equal to Julius Caesar," then he would not have taken the explicit definition to have gained anything. Yet, all he says here is that "the sense of the expression 'extension of a concept' is assumed to be known."[30] This failure to explain how recognition-statements involving terms of the form "the extension of G" are to be interpreted might be regarded as a gap in Frege's theory. And, in fact, Frege himself seems to have come to believe this, for in his later publication, *The Basic Laws of Arithmetic*,[31] where he attempted to implement in detail the plan laid out in *The Foundations of Arithmetic*, he returned to this issue and attempted to fill the gap. We will have more to say about this toward the end of this chapter. For the moment, however, we will follow Frege's original approach by assuming that extensions require no further explanation. In particular, we assume, with Frege, that *every* concept has an extension; this is an instance of the universal applicability of logic, for Frege takes the theory of extensions to be a part of logic. The existence of an extension for every concept assures us that Frege's explicit definition, (18), succeeds in assigning a number to every first-level concept.

Frege is now in a position to offer a definition of cardinal number:

(19)   *n* is a *cardinal number* if and only if there exists a concept *F* such that *n* = the number of *F*s.

This definition is not circular: "the number of *F*s" has been defined explicitly without making use of the notion of number. Note also that this definition does not define the notion of finite, or natural, number; rather, it defines the more general notion of cardinal number. Consider, for example, "the number of even natural numbers." This is a cardinal number, for it counts the elements in some collection, yet it is not equal to any natural number (0, 1, 2, and so on); it is an example of an infinite number. The next step is to do justice to this important distinction among cardinal numbers by defining "natural (or finite) number," and it is to this that Frege now turns.

First, he takes "0" to denote the number of the concept *not self-identical*. Since everything is identical to itself, nothing falls under this concept. Next, he addresses the relation of successor; he takes "succeeds" to mean "is one bigger than," and defines it as follows:[32]

(20)   The number of *F*s *succeeds* the number of *G*s if and only if there exists an *x* that falls under *F* such that the number of *G*s = the number of the concept *falls under F but is not identical to x*.

---

[30]   *Foundations*, p. 117; see also his note 1 on p. 80.
[31]   Frege (1892/1903).
[32]   *Foundations*, §76.

We can say that *n* succeeds *m* just in case there is a concept *F* such that *n* is the number of *F*s, and there is an *x* that falls under *F* such that *m* is the number of things that are *F* but not identical to *x*. Basically, *n* succeeds *m* just in case *n* is the number of a concept under which falls one more thing than there does under a concept of which *m* is the number.

What number succeeds 0 then? Frege defines 1 to be the number of the concept *identical to 0*. There is an object, 0, which falls under the concept *identical to 0*. Furthermore, by the equivalence of (16) and (17), the number of the concept *identical to 0 but not identical to 0* is the same as the number of the concept *not self-identical*, since nothing falls under either of these concepts. Therefore, by definition (20), 1 (the number of the concept *identical to 0*) succeeds 0 (the number of the concept *not self-identical*). Likewise, if we define 2 to be the number of the concept *identical to 0 or 1*, it follows from Frege's definitions that 2 succeeds 1. And it appears that we can keep going indefinitely: in terms of any defined number, we arrive at a new concept, and then we can take its number to get what can be shown to be the first number's successor.

It is worth observing here that our ability to continue generating the natural numbers *ad infinitum* by this procedure depends on the assumption that every first-level concept has a number, and furthermore that this number is an object. Once the numbers up to some number *n* have been defined, the next number is defined as the number of a certain first-level concept *F*. Of course, if the number so defined is to be the successor of *n*, the concept *F* must be chosen carefully: it must be a concept under which $n + 1$ objects fall. Frege's strategy is to define this concept *F* by specifying the $n + 1$ objects that are to fall under it, and so if his strategy is to succeed he needs a supply of $n + 1$ objects to use in the definition. It is the fact that numbers are objects that guarantees the availability of this supply: the objects he uses are the numbers from 0 to *n*, which we are assuming have already been defined and which, fortunately, are precisely $n + 1$ in number. Given the structure of Frege's procedure for generating the natural numbers, they need to be objects.

Frege's definitional victories, although infinite in number, do not yet amount to a reduction of arithmetic to logic. Taken together, they may be said to show that the individual natural numbers are logical in nature, but they do not yet say what it is about the natural numbers that makes them natural numbers, what in logical terms is distinctive of just them. What Frege needs to do next is define the concept *natural number*.

Let us first observe that successors, when they exist, are unique. Let *F* and *G* be concepts whose numbers are *y* and *z* respectively. If *y* and *z* are both successors of *x*, then there are objects *a* and *b* falling under *F* and *G* respectively such that the numbers of the concepts *falls under F but is not identical to a* and *falls under G but is not identical to b* are both *x*. But then there is a one-to-one correspondence between these last two concepts. By matching up *a* with *b*, we can extend this to a one-to-one correspondence between *F* and *G*, so the numbers of these concepts, *y* and *z*, are equal. If we use "*Sxy*" to mean that *y* is a successor of *x*, then we can state this fact as follows:

$$\forall x \forall y \forall z [(Sxy \wedge Sxz) \rightarrow y = z].^{33}$$

This legitimates our talking of *the* successor of a natural number.

We can thus say that the natural numbers are just 0, the successor of 0, the successor of the successor of 0, and so on. Our task now amounts to saying how this "and so on" is to be captured using only the resources of logic. Let us consider the concepts *F* such that (i) 0 falls under *F* and (ii) whenever *x* falls under *F* and *Sxy*, then *y* falls under *F*. Often (ii) is expressed by saying that the concept *F* is *hereditary with respect to successor*.

First Observation. If *k* is a natural number, then *k* will fall under each *F* that satisfies (i) and (ii). Why? Take any concept *F* that does satisfy (i) and (ii). Then 0 falls under *F*, by (i); hence, the successor of 0 does, by (ii); hence, the successor of the successor of 0 does, by (ii) again; and so on. We can see, then, that if *k* really is a natural number – that is, something arrived at by starting with 0 and iterating the successor relation – then it falls under *F* too. This reasoning establishes the first observation.

Second Observation. If *k* falls under every concept *F* that satisfies (i) and (ii), then *k* is a natural number. Why? Because one such *F* is precisely the concept *natural number*! So, if *k* falls under every *F* that satisfies (i) and (ii), then (since *natural number* does) *k* falls under *natural number*; and hence, of course, *k* is a natural number. The second observation is thereby established.

Our first observation states that falling under each *F* that satisfies (i) and (ii) is a necessary condition for being a natural number. Our second observation states that falling under every *F* that satisfies (i) and (ii) is a sufficient condition for being a natural number. Putting the two observations together, we get that *k* is a natural number if and only if *k* falls

---

[33] This is entailed by *Foundations*, §78, Proposition 5, which Frege states without proof.

under every concept F such that F satisfies both (i) and (ii). Expressed in logical notation:

(21) k is a *natural number* if and only if $\forall F[(F0 \wedge \forall x \forall y((Fx \wedge Sxy) \to Fy)) \to Fk]$

This is an explicit definition of *natural number* that involves nothing but 0, the successor relation, and basic logic.[34]

There are several observations about this definition that are worth making here. First, we note that the definition involves second-order quantification: it employs quantification, not just over objects, but over concepts as well.

Following on this, we observe that the adequacy of the definition depends on there being in the range of the second-order quantifier the concept *natural number* itself: the elimination of objects that are not natural numbers depends on this. For instance, why is Julius Caesar not a natural number according to this definition? Only because he does not fall under *every* concept under which 0 falls and which is hereditary with respect to successor; in particular, he does not fall under the concept *natural number* itself. Without the presence of this concept in the range of the quantifier, we could not be sure that nothing but natural numbers are counted as such by our definition. There is no circularity here: the predicate "natural number" is not used in the definition (21). Rather, the concept it denotes must be reckoned in the range of the definition's second-order quantifier.

When a definition contains a quantifier whose range includes the very entity being defined, the definition is said to be *impredicative*. The impredicativity of (21) would not have been of any concern to Frege: for him, quantifiers range over all existing entities of the appropriate level, and since the concept *natural number* exists, clearly it is in the range of any second-level quantifier. It is quite true that if someone did not have any understanding of the concept *natural number*, then (21) would not straightforwardly help him attain it. If we imagine someone attempting to apply the definition to determine whether a presented candidate is indeed a natural number, then the impredicativity of (21) guarantees that in order for this application to result in the correct outcome, he must already be acquainted with the concept *natural number* (or some other concept with the same extension) and know precisely which entities fall under it. But, as we noted earlier, Frege's project is not the psychological one of formulating definitions that might

[34]   *Foundations*, §83.

prove to be of heuristic value to mathematicians, or of pedagogical value to teachers. His interest lies rather in the logical structure of concepts, the structure by virtue of which arithmetical claims can be reduced to pure logic. This is the goal, and no other, for which the definition is enlisted by Frege, and its impredicativity is no bar to this particular goal's attainment.

That (21) satisfactorily accomplishes its task is partially confirmed by our third observation: that the validity of mathematical induction follows from the definition. To see that Frege's definition easily guarantees the validity of induction, assume that F is a concept for which the premises of mathematical induction hold; in other words, assume that 0 falls under F (the base case) and that F is hereditary with respect to successor (the induction step). If k is any natural number, then it must fall under F, since by (21) it falls under every such concept. Therefore, every natural number falls under F – just the conclusion of mathematical induction. The inference is so immediate that (21) is sometimes said to define the natural numbers as that collection of objects for which induction is valid.

Finally, we should observe that Frege's definition is in fact a particular application of a general method that can be applied to any relation Rxy to yield a new relation R^*xy that holds of two objects a, b if and only if there is a finite chain of objects $o_1, o_2, \ldots, o_n$ such that $a = o_1, b = o_n$, and $Ro_1o_2, Ro_2o_3, \ldots, Ro_{n-1}o_n$. This new relation is known as the *ancestral* of R.[35] For example, the ancestral of the relation "x is a parent of y" is just the eponymous relation "x is an ancestor of y." We can explicitly define the ancestral R^*xy in this way: $\forall F[(Fx \land \forall z \forall w((Fz \land Rzw) \rightarrow Fw)) \rightarrow Fy]$. Note that according to this definition, R^*xx always holds, so we are taking x's ancestors to include not only x's parents, x's grandparents, and so on, but also x himself. In this notation, Frege's definition says that k is a natural number if and only if S^*0k. (For more on the ancestral, see exercise 5.)

Now that Frege has defined the predicate "natural number," can he claim to have arrived at his goal of reducing arithmetic to pure logic? We have already noted that there is some question about what constitutes pure logic. And since, as we saw earlier, Frege's notion of reduction is correlative to that of his conception of logic, there is some unclarity regarding the notion of reduction as well. We shall turn to these issues in Chapter 4. For now, however, let us take for granted the notion of logical reduction, and let us consider whether there is any unclarity about what counts as a successful logical reduction *of arithmetic*. On the face of

[35]   *Foundations*, §79.

it, there is, for we do not have before us an explicit description of what such success would consist in. We know that it requires that we be able to justify all of our arithmetical knowledge on the basis of logic. But there are infinitely many arithmetical truths, and so we cannot hope to accomplish this by justifying them one by one. We need to find some fundamental arithmetical truths from which all others can be proved, and then show that they in turn can be derived from Frege's definition. In *The Foundations of Arithmetic*, Frege does not isolate any such collection of basic arithmetical truths, and so there is a gap that needs to be filled before his project's success can be evaluated.

This lacuna was in fact first remedied by Dedekind, when he advanced his axioms for arithmetic.[36] We can formulate them, now often known as Peano's Postulates,[37] as follows:

(22)   (i) 0 is a natural number.
(ii) For any x and y, if x is a natural number and y succeeds x, then y is a natural number.
(iii) Every natural number has a unique successor.
(iv) 0 is not the successor of any natural number.
(v) If x and y are natural numbers and $x \neq y$, then the successor of x is not equal to the successor of y.
(vi) For any F, if $F0$ and F is hereditary with respect to successor and k is any natural number, then Fk.

Dedekind's analysis specifies neither what 0 is, nor what the successor relation is, nor which objects the natural numbers are. A specification of the meanings of the words "0", "succeeds", and "natural number" is called a *model* for Peano's Postulates if statements (22)(i)–(vi) come out true when the words in question are given the specified meanings. Although these axioms do not single out a particular model, a particular way of interpreting the expressions that appear in them, they completely characterize the structure of any such model: in any model, the set of

---

[36] Originally published in 1888, and translated in Dedekind (1963): this translation has been revised in volume 2 of Ewald (1996), on pp. 790–833. See also Dedekind's "Letter to Keferstein," in van Heijenoort (1967), pp. 98–103. Frege himself offers axioms for arithmetic in volume 1 of Frege (1892/1903): for an analysis of Frege's axiomatization, see Heck (1995).

[37] These axioms are named after the Italian mathematician Giuseppe Peano (1858–1932), who developed axioms for arithmetic in his "The principles of arithmetic, presented by a new method" (1889), translated in van Heijenoort (1967).

objects to which the predicate "natural number" applies must consist of a first element (the one denoted by "0"), a second element (to which the first is related by the successor relation), and so on. To say this more precisely, consider any two models for Peano's Postulates. Suppose that, in the first, the set of objects to which the predicate "natural number" applies is N, the term "0" denotes some particular element a of N, and the word "succeeds" denotes a particular relation S. Suppose also that, in the second, the set of natural numbers is N', the object denoted by "0" is a', and the successor relation is S'. Although N', a', and S' need not be identical with N, a, and S, respectively, there must be a one-to-one correspondence between N and N' such that the correspondence pairs a with a', and if x and y are any two elements of N, and x' and y' are the elements of N' paired with them by the correspondence, then Sxy holds if and only if $S'x'y'$ holds. This one-to-one correspondence can be constructed by pairing a with a', pairing the unique element b of N such that Sab with the unique element b' of N' such that $S'a'b'$, and so on. Mathematicians say in this situation that the two models, although not identical, are *isomorphic*. Thus, any two models of Peano's Postulates will be isomorphic; the Postulates are therefore said to be *categorical*. The mathematical properties of the natural numbers, it is often said, are ones that relate to the structure of the numbers, rather than to the actual identities of the numbers. It follows that a claim about any such property that is true in one model for Peano's Postulates is true in all of them, and is therefore a logical consequence of the Postulates.

We can now firm up the question about whether Frege succeeds in his reduction of arithmetic to logic: for we can ask whether, given Frege's definition of the number 0, of the successor relation, and of the predicate "natural number," he can derive Peano's Postulates using only principles of logic. If so – and should we accept that these three definitions make use only of logical notions, and also that (22) captures our conception of natural number – then Frege's goal of reducing all of arithmetic to logic will have been achieved.

We have already noted that (22)(vi), a formulation of mathematical induction, follows immediately from Frege's definition of natural number. Obviously, (i) does as well, since $\forall F[(F0 \land \forall x \forall y((Fx \land Sxy) \rightarrow Fy)) \rightarrow F0]$ is a logical truth. To prove Axiom (ii), assume that x is a natural number and y succeeds x, and let F be any concept under which 0 falls and which is hereditary with respect to the successor relation: by (21), x falls under F and so, since F is hereditary with respect to successor, y does too; hence, by (21), y is a natural number. The proof of Axiom (iv) is likewise immediate: if the number of Fs succeeds some number then, directly from the definition of "succeeds," we see that something falls under *F*, and so the number of *F*s is not 0. And Axiom (v) is established by showing that Frege's successor relation is not many-one: $\forall x \forall y \forall z[(Sxz \wedge Syz) \rightarrow x = y]$ (see exercise 3).[38]

This leaves Axiom (iii), whose status as a truth of logic appears the most troublesome. We saw that for Frege one of the distinguishing features of logic is its complete generality. It does not aim to articulate truths about a particular domain of objects. Rather, its focus is on the nature of truth in general. But if this is so, why should logic entail the existence of infinitely many objects, which is what follows from (iii), in the context of the other axioms of (22)? The concern here is not that these objects are natural numbers, mathematical entities, and so not the kinds of entities to whose existence logic should be committing itself. For if their existence is entailed by logic, then they are thereby shown to be logical objects after all. We cannot rely on any intuitive conception of what is a logical object; we learn what is logical by seeing what can be reduced to the most general laws of inference. The concern is, rather, *that there should be any logical objects at all*. If logic is that which is most generally applicable, then should it not abstain from all ontological commitments, from all claims about what exists? Is its own status as logic not thrown into doubt by the existential assumptions it embodies?

This concern is not decisive. It seems to have force because it is assumed that a world in which no objects exist, or one in which only finitely many objects exist, is a possible world. Given this assumption, logic's apparent infinitary ontological commitments of course conflict with its pretension to be universal in application. But the assumption is only that. It may seem as if there is no logical incoherence in imagining a world with no objects, just as it may have once seemed to be coherent to imagine that $\sqrt{2}$ is rational. But our intuitions and imaginings are as may be. We now know that there could not be a world in which $\sqrt{2}$ is rational, or one in which there are only finitely many prime numbers. Likewise, if the laws of thought entail that there are infinitely many logical objects, then the possibility that there might not be is thereby shown to be illusory and ultimately unintelligible.

Now we must see how Frege establishes Axiom (iii).[39] The proof makes use of the ancestral S^*xy of the successor relation Sxy. Recall that, by definition, S^*xy holds if and only if $\forall F[(Fx \wedge \forall z \forall w((Fz \wedge Szw) \rightarrow Fw)) \rightarrow Fy]$. The intention here is that, for natural numbers x and y, S^*xy will hold if and only if $x \leq y$.

[38] This is entailed by *Foundations*, §78, Proposition 5.
[39] *Foundations*, §82.

It is not hard to show that $S*00$ is true, but for any $x \neq 0$, $S*x0$ is false. Another way to say this is that the predicate "$S*\_\_0$" denotes a concept, and the only object that falls under this concept is 0. It follows that the number of objects falling under the concept denoted by "$S*\_\_0$" is 1, which is the successor of 0. Similarly, the only objects that fall under the concept denoted by "$S*\_\_1$" are 0 and 1, and so the number of objects falling under this concept is 2, the successor of 1. This reasoning is of course similar to that which motivated Frege's definitions of 1 and 2.

These examples provide the motivation for Frege's proof of Axiom (iii). Frege proves, by mathematical induction, that, for every natural number n, the number of objects falling under the concept denoted by "$S*\_\_n$" is a successor of n; since we earlier established that successors, when they exist, are unique, it follows that every natural number n has a unique successor. We have already observed that the number of objects falling under the concept denoted by "$S*\_\_0$" is the successor of 0, which takes care of the base case of the induction. For the induction step, suppose that m is the successor of n, and the number of objects falling under the concept denoted by "$S*\_\_n$" is m. We must show that the number of objects falling under the concept denoted by "$S*\_\_m$" is the successor of m. We will not go through the details; the key idea, however, is to show that the objects that fall under the concept denoted by "$S*\_\_m$" are those that fall under the concept denoted by "$S*\_\_n$", together with m. The desired conclusion then follows easily from Frege's definition of the successor relation (see exercise 9).

This proof of Axiom (iii), like Frege's definitions of 1 and 2 and his procedure for continuing to generate larger natural numbers, relies on the assumption that every first-level concept has a number. And, as we observed earlier, this assumption can be justified if we can be assured that every concept has an extension. Thus, the question of whether or not the existence of successors for all natural numbers has been demonstrated by means of logic alone ultimately comes down to the question of whether or not the assertion that all concepts have extensions is a logical truth.

Of course, there is more to mathematics than arithmetic. Contemporary mathematicians make use not just of natural numbers, but of the infinite collection of all natural numbers, along with many other infinite sets. We shall explore this in some detail in the following chapter. It suffices to note here that Frege already has the resources to begin this stage of the reduction. For recall that he has defined the concept *natural number*. He can therefore speak of the extension of that concept, which will be infinite. This seems intuitively obvious as 0, 1, 2, and infinitely many other objects fall under the concept *natural number*. But can Frege prove it? He can, by showing that the number of natural numbers is not identical to any natural number. This is so because, while no natural number is identical to its successor, the number of natural numbers is! (See exercise 10.) Again, we will return to these issues in Chapter 3. It bears observing that a general theory of extensions will play an important part in these further developments as well. It is because Frege assumed that every concept has an extension that he could confidently refer to the number of natural numbers. Extensions will also play a major role in the development of the real numbers.

It is no surprise, then, that Frege felt the need to treat of extensions in far greater detail in his *The Basic Laws of Arithmetic*. In this work, Frege attempted to explain the meaning of the expression "the extension of F," for any concept F, by a method similar to the contextual definition method that he tried and then abandoned in his earlier definition of the cardinal numbers. An important part of this analysis was his Basic Law (V), according to which, for any concepts F and G, the extension of F is equal to the extension of G if and only if precisely the same objects fall under F and G. (Note the similarity of this rule to the equivalence between (16) and (17), which was Frege's partial contextual definition of "the number of Fs.") For a time he believed that he had succeeded in eliminating any doubt about the logical nature of extensions by demonstrating, by completely general means, the existence of extensions for all concepts. The goal of a reduction of arithmetic to logic appeared to be within sight. And then disaster struck. For, as we will see shortly, his method for establishing the existence of extensions for all concepts was not – indeed, could not possibly have been – correct. In the following chapter, we will discuss the catastrophe that befell Frege's project of reducing arithmetic, and eventually all of analysis, to a completely general theory. We will also present in some detail a modern approach to its execution that, so far as we know, steers clear of the paradox that beset Frege's system. Then, in Chapter 4, we will consider whether this approach would have fulfilled Frege's philosophical ambitions.

Exercises

1. Show that the relation of equinumerosity has the following properties:
    (a) For any concept F, F is equinumerous with F.
    (b) For any concepts F and G, if F is equinumerous with G, then G is equinumerous with F.
    (c) For any concepts F, G, and H, if F is equinumerous with G and G is equinumerous with H, then F is equinumerous with H.

2. Show that Frege's partial contextual definition of "the number of *Fs*" follows from his explicit definition. In other words, use the explicit definition to prove that, for any concepts *F* and *G*, the number of *Fs* = the number of *Gs* if and only if *F* is equinumerous with *G*.

3. Show that, for any numbers *a*, *b*, and *c*, if *Sac* and *Sbc*, then *a* = *b*. This establishes Peano Postulate (22)(v). (Hint: Suppose that *a*, *b*, and *c* are the numbers of concepts *F*, *G*, and *H* respectively. By the definition of the successor relation, there must be objects *p* and *q* falling under *H* such that *F* is equinumerous with the concept *falls under H but is not identical to p*, and *G* is equinumerous with the concept *falls under H but is not identical to q*. Now show, by matching up *q* with *p*, that the concepts *falls under H but is not identical to p* and *falls under H but is not identical to q* are equinumerous.)

4. Prove that, for every natural number *n*, ¬*Snn*. (Hint: Use mathematical induction. In the induction step, you may find it helpful to apply exercise 3.)

5. Prove that the ancestral *R** of a relation *R* has the following properties:
   (a) For all *a* and *b*, if *Rab*, then *R*ab*.
   (b) For every *a*, *R*aa*.
   (c) For all *a*, *b*, and *c*, if *R*ab* and *R*bc*, then *R*ac*.

6. Prove that, for every number *a*, if *S*a0*, then *a* = 0. (Hint: Suppose *S*a0* but *a* ≠ 0. Let "*Fx*" mean "*x* ≠ 0"; in other words, let *F* be a concept under which falls everything except 0. Now show that *Fa* and *F* is hereditary with respect to successor, but ¬*F0*, contrary to the definition of *S*a0*.)

7. Prove that, for all numbers *n* and *m*, if *Snm*, then $\forall v((S^*vm \land v \neq m) \rightarrow S^*vn)$. (Hint: Assume *Snm*, *S*vm*, and *v* ≠ *m*. To prove *S*vn*, let *F* be a concept such that *Fv* and *F* is hereditary with respect to successor; we must prove *Fn*. Assume ¬*Fn*. Let "*Gx*" mean "*Fx* ∧ *x* ≠ *m*." Now verify that *Gv* and *G* is hereditary with respect to successor, but ¬*Gm*, contradicting the assumption that *S*vm*.)

8. Prove that, for all natural numbers *n* and *m*, if *Snm*, then $\forall v(S^*vn \rightarrow (S^*vm \land v \neq m))$. (Hint: Use induction on *n*. In other words, for the base case, assume that *S0m* and prove that $\forall v(S^*v0 \rightarrow (S^*vm \land v \neq m))$; you will find exercises 4, 5, and 6 helpful for this. Then, for the induction step, assume that *Snm*, *Smp*, and $\forall v(S^*vn \rightarrow (S^*vm \land v \neq m))$, and prove that $\forall v(S^*vm \rightarrow (S^*vp \land v \neq p))$. You will find exercises 4, 5, and 7 helpful for this.)

9. Prove Peano Postulate (22)(iii): every natural number has a successor. (Hint: Prove by mathematical induction that, for every natural

number n, the numbers of x's such that S^*xn is the successor of n. For the base case, apply exercises 5 and 6. For the induction step, combine exercises 7 and 8 to conclude that if Snm, then the number of x's such that S^*xm is the successor of the number of x's such that S^*xn.)

10. Let a be the number of the concept *natural number*. Prove that Saa. Notice that it follows, by exercise 4, that the number of natural numbers is not itself a natural number. (Hint: Show that the successor relation S is a one-to-one correspondence between the concept *natural number* and the concept *natural number not identical to 0*.)

3

Set Theory

As we saw in the last chapter, in his attempt to base mathematics on logic, Frege relied on the assumption that every concept has an extension. It was therefore a tremendous blow to Frege's program when Bertrand Russell discovered in 1901 that this assumption leads to a contradiction. The contradiction is known today as Russell's Paradox. Russell informed Frege of the contradiction in a letter dated June 16, 1902, shortly before the publication of volume 2 of Frege's *The Basic Laws of Arithmetic*. Frege immediately recognized that his analysis of extensions must be flawed, and in particular that his Basic Law (V) could not hold in complete generality. In a hastily added appendix to volume 2 he proposed a restriction on Basic Law (V) but, as he eventually realized, the proposal failed to solve the problem. The planned volume 3 was never published.

To see how Russell's Paradox is derived, we must first note that some concepts apply to sets, and the extensions of such concepts are sets that contain sets as elements. For example, consider the concept *finite set*. Its extension F is the set of all sets with finitely many elements. In other words,

(1) $F = \{x : x \text{ is a finite set}\}$.

This is read "F is the set of all x such that x is a finite set." The elements of F include, for example, the set of all Plagues of Egypt and the set of all stars in our galaxy, but not the set of all natural numbers. This last set falls under the concept *infinite set*, and is therefore an element of its extension, the set

(2) $I = \{x : x \text{ is an infinite set}\}$.

Although the elements of F are finite sets, F itself is an infinite set, since it includes among its members all of the sets in the infinite list

$\{0\}, \{1\}, \{2\}, \ldots$. Thus, F is not a member of itself but, rather, is a member of I. In symbols, we write $F \notin F$ and $F \in I$, respectively. Similarly, the members of I include the infinitely many sets $\{0, 1, 2, \ldots\}, \{1, 2, 3, \ldots\}, \{2, 3, 4, \ldots\}, \ldots$, so I is also an infinite set, and therefore $I \in I$. The sets F and I illustrate that, according to the conception of sets that Frege was using, some sets are members of themselves, and some are not. Russell proposed that we consider the concept *set that is not a member of itself*. Let R be the extension of this concept:

(3) $R = \{x : x \text{ is a set and } x \notin x\}$.

Then the discussion above shows that $F \in R$ but $I \notin R$. But now suppose we ask whether or not $R \in R$. According to the definition of R, R will be a member of R if and only if it falls under the concept *set that is not a member of itself*. In other words,

(4) $R \in R$ if and only if $R \notin R$.

But this is impossible: whether R is a member of itself or not, statement (4) cannot be true. Thus, the assumption that the concept *set that is not an element of itself* has an extension has led to a contradiction.[1]

Many people find Russell's Paradox surprising, because it is difficult to see what is wrong with defining R as the extension of the concept *set that is not an element of itself*. This concept seems to be specified by an unambiguous criterion that applies to some sets but not others. Why can we not sort all sets into those to which the criterion applies and those to which it does not, and then collect together the sets to which it applies to form the set R? But the claim that R can be formed does not lead immediately to Russell's Paradox. The paradox results only if we assume that R is one of the sets that are to be sorted using the concept *set that is not an element of itself*. In other words, to produce the paradox we must view R as an object that *already exists*, even before we carry out the sorting of sets that would be involved in forming R.

Thus, Russell's Paradox can be thought of as resulting from the assumption that if it is possible to collect some objects together to form a

---

[1] To see how the derivation of Russell's Paradox is carried out using the formal rules of Frege's *The Basic Laws of Arithmetic*, and in particular to see the role played in the derivation by Basic Law (V), the reader can consult Boolos (1986/ 87, 173).

set, then the collecting is actually unnecessary; the set containing those objects already exists. Such an assumption is natural from the point of view of platonism, according to which mathematical objects exist independently of us and our activities, and hence independently of any collecting of objects that we might carry out. This assumption leads to a circularity in which R plays a role in its own formation, and this circularity then leads to the paradox. Russell suggested that in order to avoid this paradox (and other related paradoxes) such circularities must be eliminated, and he proposed a rule, which he called the "vicious-circle principle," to ban them: "Whatever involves *all* of a collection must not be one of the collection."[2] In collaboration with the American philosopher Alfred North Whitehead (1861–1947), he set out to revise and extend Frege's work by basing mathematics on a version of logic that incorporates the vicious-circle principle, and therefore avoids the paradoxes.

Russell and Whitehead presented their theory, which is known as the *ramified theory of types*, in the three-volume work *Principia Mathematica*, published during the years 1910–13. To explain their theory, it may be easier to begin with the simpler *unramified theory of types*. In the unramified theory of types, all objects are classified into a hierarchy of types. The lowest level of the hierarchy consists of individual objects that are not sets. Next come sets whose elements are individuals, and then sets of sets of individuals, and so on. Variables in the theory cannot range over all objects, but must range over only objects of some particular type.

For example, consider a set A, defined as the set of all x for which some statement $P(x)$ is true:

(5) $\quad A = \{x : P(x)\}.$

The theory of types requires that the variable x in this definition be restricted to range over objects of some particular type, and the type of the set A being defined will then be one level higher in the type hierarchy. For example, if x ranges over individuals, then the set A will be a set of individuals; if x ranges over sets of individuals, then A will be a set of sets of individuals. Thus, A is not a possible value of x, so A cannot be considered for elementhood in itself. It follows that, in the derivation of Russell's Paradox, the step from (3) to (4) is blocked, and therefore the paradox is avoided. In fact, in the theory of types the definition (3) is not

---

[2] "Mathematical logic as based on the theory of types," reprinted in van Heijenoort (1967, 150–82). The quotation is from p. 155.

even allowed, since x is not the right type of entity to be an element of itself, so the expression "$x \notin x$" that appears in (3) is ill-formed.

Russell and Whitehead's ramified theory of types is somewhat more complicated, because they worked not with sets but, rather, with what they called *propositional functions*, which are similar to Frege's concepts. Furthermore, Russell and Whitehead's rules concerning the assignment of types to variables are more restrictive than the rules of the unramified theory. According to the unramified theory, in definition (5) the type of the variable x, which occurs as a free variable in the statement $P(x)$, must be lower than the type of the set A being defined. But Russell and Whitehead insisted that not only free variables but also bound variables used in the definition of a propositional function must have types that are lower than the type of the propositional function being defined. Thus, in particular, a definition of a propositional function cannot involve quantification over a collection that includes the propositional function being defined. In other words, impredicatively defined propositional functions are ruled out. This restriction resulted in a further proliferation of categories in their hierarchy of propositional functions.

This ban on impredicative definitions prevented Russell and Whitehead from deriving all of mathematics in their type theory as we have described it so far, because a number of important definitions in mathematics are impredicative. For example, as we saw in the last chapter, Frege's definition of the natural numbers is impredicative. For this reason, Russell and Whitehead found it necessary to include in their logical system an axiom that they called the Axiom of Reducibility. Without going into detail about exactly what this axiom says, we will simply note here that it had the effect of making impredicative definitions possible.

One feature of the theory of types that might be regarded as unfortunate is that it rules out not only sets that lead to paradox, such as Russell's set, but also other sets that seem harmless. For example, suppose that a is an individual and b is a set of individuals. Then in the theory of types we cannot talk about the set $\{a, b\}$, since it is not an individual, or a set of individuals, or a set of sets of individuals, or any other type of object allowed in the theory of types. But it is hard to see what harm can come from putting a and b together in the same collection.

A second problem with the theory of types is that if there are only finitely many individuals, then there will be only finitely many sets of individuals, finitely many sets of sets of individuals, and so on. Thus, although the entire type hierarchy will be infinite, each level of the hierarchy will be finite. Since each variable must range over only objects of a particular type, this means that no variable can range over infinitely many objects. For this reason, Russell and Whitehead found that in order to develop mathematics in type theory they needed to include an axiom asserting that there are infinitely many individuals. But, as we observed in the previous chapter, one might question whether an axiom that makes such ontological commitments should be regarded as a principle of logic.

Today, most mathematicians prefer to use a theory of sets that avoids the paradoxes without requiring a classification of all entities into types, without prohibiting impredicative definitions, and without making assumptions about the existence of anything other than sets. The most popular such theory is the Zermelo–Fraenkel theory of sets, usually abbreviated ZFC. (The "C" stands for the Axiom of Choice, a somewhat controversial axiom that will be discussed later in this chapter.) Most of the ZFC axioms were formulated by the German mathematician Ernst Zermelo (1871–1953) in 1908,[3] building on earlier ideas of the mathematician Georg Cantor (1845–1918).[4] In 1922, the German mathematician Abraham Fraenkel (1891–1965) clarified and extended Zermelo's system, bringing it into its modern form. (Similar improvements were proposed independently at about the same time by the Norwegian mathematician Thoralf Skølem (1887–1963).) In the rest of this chapter we will introduce ZFC and see how it can be used as a foundation for mathematics.

The theories developed by Frege, and also Russell and Whitehead, were intended to show how logic could be applied to statements about anything at all. In contrast, ZFC is a theory that is concerned only with sets. Even the elements of all of the sets studied in ZFC are also sets (whose elements are also sets, and so on). It might be thought that the study of such "pure" sets would be faced immediately with an infinite regress: If all the elements of a pure set are also pure sets, and all their elements are pure sets, etc., then how can we ever get started writing down pure sets? However, it is easy to name one pure set: the empty set, denoted $\varnothing$. The empty set contains nothing that is not a pure set, because it contains nothing at all! But now, using $\varnothing$ as an element, we can write down another pure set, $\{\varnothing\}$. Note that $\{\varnothing\}$ is different from $\varnothing$; the latter has

---

[3]   "Investigations in the foundations of set theory I," reprinted in van Heijenoort (1967, 199–215).

[4]   Cantor was born in Russia but was raised and lived in Germany. His work on set theory was motivated by his study of trigonometric series, and the sets of real numbers he found it necessary to work with in this study. For more on the history of set theory, see Dauben (1979) and Hallett (1984).

no elements, while the former has one element, namely $\emptyset$. From these two sets we can construct more pure sets, such as $\{\emptyset, \{\emptyset\}\}$ and $\{\{\emptyset\}\}$. ZFC is intended to be the theory of these and similar sets. For the rest of this chapter, when we say "set" we mean "pure set."[5]

ZFC is a formal axiomatic theory. By this we mean that there is a *formal language*, which we will call *the language of set theory*, in which all our statements about sets can be written; that there are certain statements in this formal language that are called *axioms* for the theory; and that there are *formal rules of inference* by which all theorems of ZFC can be derived from these axioms.

The language of set theory uses the usual symbols of logic ("$\wedge$", "$\vee$", "$\neg$", "$\rightarrow$", "$\leftrightarrow$", "$\forall$", "$\exists$", "(", ")", and "=") together with the symbol "$\in$" (meaning "is an element of") and variables (which always stand for sets). Here are a few examples of meaningful expressions in the language of set theory:

Expression	Meaning
$\exists x \forall y \neg(y \in x)$	There is a set that has no elements; that is, the empty set exists.
$\forall z(z \in x \rightarrow z \in y)$	Every element of x is an element of y. This is usually expressed by saying that x is a *subset* of y, or in symbols, $x \subseteq y$.

The axioms of ZFC consist of the usual axioms of first-order logic, together with certain statements about sets whose truth is considered by most mathematicians to be intuitively clear. The rules of inference are simply the usual rules of inference of first-order logic.

All that remains to specify the theory ZFC is to list the set-theoretic axioms of the theory. The first of these axioms states one of the most fundamental properties of sets, the fact that a set is determined by its elements. In other words, if two sets have exactly the same elements, then they must be equal:

---

[5] It is possible to modify ZFC to construct a theory that allows for the existence of nonsets, and of sets other than pure sets. Such a modified theory would be more similar to the theories developed by Frege and by Russell and Whitehead, and might be more useful for explaining how mathematics can be applied to phenomena in the real world. However, for the purpose of developing pure mathematics, mathematicians have found that nothing is needed other than pure sets, and it has therefore become customary in the study of set theory to study only pure sets. We have chosen to be consistent with this tradition by presenting ZFC in its usual form.

AXIOM OF EXTENSIONALITY.   For all sets x and y, if x and y have the same elements then $x = y$. Or, in the language of set theory:
$$\forall x \forall y \ (\forall z(z \in x \leftrightarrow z \in y) \rightarrow x = y).$$

For example, suppose A is the set of all solutions to the equation $x^2 = 2x$, and B is the set of all solutions to the equation $(x - 1)^2 = 1$. Then, by solving these equations you can show that A and B have exactly the same elements, namely the numbers 0 and 2. Therefore, by the Axiom of Extensionality, $A = B$.

Before introducing more of the axioms of ZFC, it will be helpful to describe in a bit more detail the universe of sets these axioms are intended to describe. To avoid vicious circles that might lead to paradoxes, we will imagine that these sets are created in stages, with sets created at one stage containing as elements only sets that were created at earlier stages. Of course, this idea is reminiscent of the hierarchy of sets in the theory of types. We will let V_n denote the collection of all sets constructed by stage n. Initially, no sets have been constructed, so we let:

(6) $V_0 = \varnothing$.

To describe the sets constructed at later stages the following terminology will be helpful:

DEFINITION 3.1.   The set of all subsets of A is called the *power set* of A, and is denoted $\mathcal{P}(A)$. In other words,

(7) $\mathcal{P}(A) = \{B : B \subseteq A\}$.

For example, if $A = \{1, 2\}$ then A has four subsets, namely $\varnothing, \{1\}, \{2\}$, and $\{1, 2\}$. Therefore $\mathcal{P}(A) = \{\varnothing, \{1\}, \{2\}, \{1, 2\}\}$. The relevance of this idea to our description of the universe of sets is that the power set of A can be thought of as the collection of all sets that can be constructed using members of A as elements.

Returning now to our construction of sets in stages, recall that V_n is the collection of all sets constructed by stage n, and when constructing a set at the next stage we may use only sets in V_n as elements. In other words, every set constructed at the next stage must be a subset of V_n. In fact, we will assume that every subset of V_n gets constructed at the next stage (if it has not been constructed already). Thus, the collection of sets constructed by stage $n + 1$ will consist of all the sets in V_n, together with all subsets of V_n:

(8) $V_{n+1} = V_n \cup \mathcal{P}(V_n)$.

Note that this is a recursive definition, so we can use it to work out V_1, V_2, and so on, in order. For example:

(9) $V_1 = V_0 \cup \mathcal{P}(V_0) = \varnothing \cup \mathcal{P}(\varnothing) = \varnothing \cup \{\varnothing\} = \{\varnothing\},$
$V_2 = V_1 \cup \mathcal{P}(V_1) = \{\varnothing\} \cup \mathcal{P}(\{\varnothing\}) = \{\varnothing\} \cup \{\varnothing, \{\varnothing\}\}$
$\quad = \{\varnothing, \{\varnothing\}\},$
$V_3 = V_2 \cup \mathcal{P}(V_2) = \{\varnothing, \{\varnothing\}\} \cup \mathcal{P}(\{\varnothing, \{\varnothing\}\})$
$\quad = \{\varnothing, \{\varnothing\}\} \cup \{\varnothing, \{\varnothing\}, \{\{\varnothing\}\}, \{\varnothing, \{\varnothing\}\}\}$
$\quad = \{\varnothing, \{\varnothing\}, \{\{\varnothing\}\}, \{\varnothing, \{\varnothing\}\}\},$
and so on.

(The observant reader will have noticed that in every case so far we have $V_n \subseteq \mathcal{P}(V_n)$, so (8) could be replaced with the simpler equation $V_{n+1} = \mathcal{P}(V_n)$. In fact, it is not hard to prove by mathematical induction that this pattern will continue.)

All of the sets constructed at all of the stages described so far are finite sets. But mathematics needs infinite sets too, so we cannot stop yet. To see how the construction of sets can be continued, let us first gather together all of the sets constructed so far into one collection, called V_ω. (For our purposes, the subscript ω here can be thought of as simply a label to distinguish this stage from all the others.)

(10) $V_\omega = V_0 \cup V_1 \cup V_2 \cup \ldots .$

To continue the construction we simply proceed as before, forming new sets using the sets in V_ω as elements:

(11) $V_{\omega+1} = V_\omega \cup \mathcal{P}(V_\omega),$
$V_{\omega+2} = V_{\omega+1} \cup \mathcal{P}(V_{\omega+1}),$
and so on.

The first infinite sets appear in $V_{\omega+1}$. For example, V_ω contains all of the sets $\varnothing, \{\varnothing\}, \{\{\varnothing\}\}$, and so on, so $V_{\omega+1}$ contains the infinite set $\{\varnothing, \{\varnothing\}, \{\{\varnothing\}\}, \ldots\}$.

Of course, we are still not done, since we can construct still more sets:

(12) $V_{\omega+\omega} = V_0 \cup V_1 \cup V_2 \cup \ldots \cup V_\omega \cup V_{\omega+1} \cup V_{\omega+2} \cup \ldots,$
$V_{\omega+\omega+1} = V_{\omega+\omega} \cup \mathcal{P}(V_{\omega+\omega}),$
$V_{\omega+\omega+2} = V_{\omega+\omega+1} \cup \mathcal{P}(V_{\omega+\omega+1}),$
and so on.

It is difficult to say how long this process is to continue. To say that it goes on "forever" is inadequate, since by the time we reached V_ω the process had already gone on forever, and yet we continued. If this makes our description of the process of constructing sets seem vague to you, do not worry; our purpose in describing the construction of sets is merely to motivate the remaining axioms of ZFC. The axioms of ZFC will be statements about sets that we will be able to see are true, even based on our somewhat vague description of the universe of sets. There will be no vagueness in the axioms themselves, and, as described earlier, the entire theory will then be derived from the axioms.

We are ready now to return to listing axioms of ZFC. Most of the remaining axioms assert the existence of sets formed from given sets in certain ways. For example, our next axiom says that given any two sets x and y, we can form the set $\{x, y\}$ whose elements are just the two sets x and y:

AXIOM OF PAIRING.   For all sets x and y, there is a set z containing x and y, and nothing else. In symbols: $\forall x \forall y \exists z \forall w (w \in z \leftrightarrow (w = x \lor w = y))$.

Note that if $x = y$, then the set z whose existence is assured by the Axiom of Pairing contains just x and nothing else; that is, $z = \{x\}$.

To see that the Axiom of Pairing is true of the universe of sets described earlier, suppose we have two sets x and y. Each of these sets must have been constructed at some stage, so there must be some stage α by which both x and y have been constructed. Therefore $x, y \in V_\alpha$, so $\{x, y\}$ is a subset of V_α and will be constructed at the next stage.

Our next axiom says that given a set x, we can form the power set of x. To see why this is true, note that if x was constructed at stage α, then since x can contain only previously constructed sets, all elements of x must have been constructed before stage α. But then every subset of x would also be constructed at stage α (if not earlier), so all of these subsets could be collected together to form the power set of x at stage $\alpha + 1$.

AXIOM OF POWER SETS.   For every set x, there is a set y whose elements are all the subsets of x. In symbols: $\forall x \exists y \forall z (z \in y \leftrightarrow \forall w (w \in z \rightarrow w \in x))$.

Another basic set-theoretic operation is the formation of unions of sets. If x is any set, then by $\cup x$ we mean the union of all the sets in x. For example, if $x = \{a, b, c, d\}$, then $\cup x = a \cup b \cup c \cup d$. More precisely, the elements of $\cup x$ are those sets that are elements of at least one element of x. To see why $\cup x$ is a set, note that if x was constructed at stage α then all its elements were constructed earlier, so all of *their* elements – that is, all elements of ∪x – were constructed even earlier. Therefore ∪x could also be constructed at stage α.

AXIOM OF UNIONS. For every set x, ∪x is also a set. In symbols: $\forall x \exists y \forall z (z \in y \leftrightarrow \exists w (w \in x \,\&\, z \in w))$.

As an example of how the axioms are used to prove theorems, let us prove that given any three sets there is a set containing those three sets and nothing else. (This explains why we have not introduced, in addition to the Axiom of Pairing, an "Axiom of Tripling.") We will present the proof in English, but readers experienced in formal logic should be able to translate our proof into a derivation written entirely in the language of set theory, in which all inference steps are justified by rules of logic.

THEOREM 3.2. *For any sets x, y, and z, there is a set w containing x, y, and z, and nothing else.*

Proof. By the Axiom of Pairing, we can let $u = \{x, y\}$, $v = \{z\}$, and $t = \{u, v\}$. But then by the Axiom of Unions we can form the set $w = \bigcup t = u \cup v = \{x, y\} \cup \{z\} = \{x, y, z\}$, as required. ∎

What about Frege's assumption that every concept has an extension? If $P(x)$ is a statement with x as a free variable, will the set $\{x : P(x)\}$ ever be constructed? For any stage α we could pick out those sets *constructed by stage* α which, when substituted for x, make the statement $P(x)$ true, and we could gather them into a set at the next stage. In other words, we could form the set $z = \{x \in V_\alpha : P(x)\}$, which is a subset of V_α and therefore an element of $V_{\alpha+1}$. But there might be sets constructed later which, when substituted for x, make $P(x)$ come out true, and these sets would not be included in z. More generally, we might want to define a set using a statement $P(x, w_1, w_2, \ldots, w_n)$ that relates x to some other sets $w_1, w_2, \ldots, w_n$. For any sets $y, w_1, w_2, \ldots, w_n$, once y has been constructed, we will also be able to construct $\{x \in y : P(x, w_1, w_2, \ldots, w_n)\}$, but there might never be a stage at which we can construct $\{x : P(x, w_1, w_2, \ldots, w_n)\}$. For this reason, ZFC includes only a restricted version of Frege's fundamental assumption.

AXIOM OF COMPREHENSION. For any formula $P(x, w_1, w_2, \ldots, w_n)$ in the language of set theory whose only free variables are $x, w_1, w_2, \ldots, w_n$, any choice of sets for the variables $w_1, w_2, \ldots, w_n$ to stand for, and any set y, there is a set containing precisely those elements of y

which, when substituted for x, make the formula $P(x, w_1, w_2, \ldots, w_n)$ true. In symbols: for every formula $P(x, w_1, w_2, \ldots, w_n)$ in the language of set theory whose only free variables are $x, w_1, w_2, \ldots, w_n$, the following statement is an axiom of ZFC:

$$(13) \quad \forall w_1 \forall w_2 \ldots \forall w_n \forall y \exists z \forall x (x \in z \leftrightarrow (x \in y \land P(x, w_1, w_2, \ldots, w_n))).$$

The Axiom of Comprehension is actually what is called an *axiom scheme*. This means that it is an infinite list of axioms, one for each formula $P(x, w_1, w_2, \ldots, w_n)$ in the language of set theory. For example, taking $P(x, w_1, w_2)$ to be the formula "$\neg(w_1 \in x \land w_2 \in x)$", we see that the following statement is an axiom of ZFC:

$$(14) \quad \forall w_1 \forall w_2 \forall y \exists z \forall x (x \in z \leftrightarrow (x \in y \land \neg(w_1 \in x \land w_2 \in x))).$$

This axiom says that for any sets w_1, w_2, and y we can form the set $z = \{x \in y : x \text{ does not contain both } w_1 \text{ and } w_2 \text{ as elements}\}$. Similarly, each choice for $P(x, w_1, w_2, \ldots, w_n)$ leads to another axiom of ZFC. If we were using Frege's higher-order logic, then we could write the Axiom of Comprehension as a single axiom starting with "$\forall P$". But the logic of ZFC is first-order, so such quantification over concepts is not allowed and therefore an axiom scheme is needed.

It is worth noting that the Axiom of Comprehension allows impredicative definitions. That is, the statement $P(x, w_1, w_2, \ldots, w_n)$ may involve quantification over all sets, a collection that includes the set z being defined. Acceptance of the Axiom of Comprehension depends on the assumption that, at any stage α in the construction of the universe of sets, *all possible* subsets of V_α are included in $V_{\alpha+1}$. In other words, it is assumed that if it is possible to collect together certain elements of V_α, even if this is done by means of a definition that makes reference to sets not yet constructed at that stage, then that collection of elements of V_α was included in $V_{\alpha+1}$.

This assumption is sufficiently similar to the assumption that led to Russell's Paradox that the reader might wonder if Russell's Paradox can be derived in ZFC. The Axiom of Comprehension would not justify the formation of Russell's set $R = \{x : x \notin x\}$, but for any set A the Axiom of Comprehension does allow us to form the set $R_A = \{x \in A : x \notin x\}$. Imitating the derivation of Russell's Paradox, suppose we ask now whether or not $R_A \in R_A$. If $R_A \in A$ then, as before, we can conclude that $R_A \in R_A \leftrightarrow R_A \notin R_A$, which is impossible. But this is no longer a paradox; it is simply a proof, by *reductio ad absurdum*, that $R_A \notin A$. Since this procedure could be used for any set A, we have shown that for every set *A* there is a set that is not an element of *A*. In other words, Russell's Paradox has been transformed into a proof of the following theorem:

THEOREM 3.3. *There is no universal set; that is, there is no set that contains all sets as elements.*

It is important to recognize that Theorem 3.3 does not say that there is no collection of all sets; in fact, in accepting the axioms of ZFC as true statements we are implicitly assuming that there is such a collection, and we are using it as the universe over which all variables in the language of set theory range. What Theorem 3.3 says is that the collection of all sets is not itself a set. Thus the axioms of ZFC force us to recognize that not every collection of sets is a set. Set theorists call collections of sets *classes*, and collections of sets that are not themselves sets are called *proper classes*. Thus, the collection of all sets is a proper class. Note that, since the variables in the language of set theory range over the universe of sets, variables always stand for sets and not proper classes.

We can get a better understanding of this phenomenon by returning to our conception of sets as being constructed in stages. According to this conception, the word "set" is reserved for collections of sets *that are constructed at some stage*. The reason the collection of all sets is not a set is that there is no stage at which the collection of all sets could be constructed. One could, at any stage, construct the collection of all sets constructed *so far*, but there would always be sets constructed at later stages that would not be in this collection. Thus the collection of all sets can never be constructed, so it cannot be a set.

Although it appears that Russell's Paradox cannot be derived in ZFC, this is not a guarantee that the theory is paradox-free. It would certainly be desirable to have a proof that no contradictions can be derived in ZFC; that is, a proof that ZFC is *consistent*. We will have much more to say about the possibility of such a consistency proof in Chapters 6 and 7. For now, we will simply observe that, although no contradictions in ZFC have ever been found, we have no guarantee that such contradictions will not arise in the future.

There are several other axioms of ZFC, but the list we have given so far is enough to begin the work of deriving all of mathematics in ZFC. We will introduce a few of the other axioms when the need for them arises.

Let us see now if we can imitate Frege's approach to defining the natural numbers, justifying all of our reasoning from the axioms of ZFC. Fundamental to Frege's definition is the idea of a one-to-one correspondence, which involves *pairing* the objects in one collection with those in another. To express this in the language of ZFC we will need the notion of an *ordered pair* $\langle a, b \rangle$. Since the only objects discussed in ZFC are sets, we will need to define $\langle a, b \rangle$ to be some set defined in terms of a and b. Note that we cannot use the set $\{a, b\}$ whose existence is guaranteed by the Axiom of Pairing, because it does not indicate the ordering of the coordinates a and b; $\{a, b\}$ and $\{b, a\}$ are the same set (by the Axiom of Extensionality), but if $a \neq b$ then we must have $\langle a, b \rangle \neq \langle b, a \rangle$. There are a number of ways of adding additional data to the unordered pair $\{a, b\}$ to incorporate this ordering information. The one used most often today was suggested by the Polish logician Kazimierz Kuratowski (1896–1980) in 1921:

DEFINITION 3.4. For any sets a and b, the *ordered pair* $\langle a, b \rangle$ is the set $\{\{a, b\}, \{a\}\}$. (Note that the existence of this set is guaranteed by three applications of the Axiom of Pairing.)

It is not hard to show that, for any sets a, b, c, and d, $\langle a, b \rangle = \langle c, d \rangle$ if and only if $a = c$ and $b = d$.

For any two sets A and B, the set of all ordered pairs $\langle a, b \rangle$ in which $a \in A$ and $b \in B$ is called the *Cartesian product* of A and B, and is denoted $A \times B$. In other words,

(15) $A \times B = \{x : \exists a \exists b (a \in A \land b \in B \land x = \langle a, b \rangle)\}$.

For example, if $A = \{p, q\}$ and $B = \{x, y, z\}$, then $A \times B = \{\langle p, x \rangle, \langle p, y \rangle, \langle p, z \rangle, \langle q, x \rangle, \langle q, y \rangle, \langle q, z \rangle\}$.

But do the axioms of ZFC guarantee that for all sets A and B there is a set fitting our description of $A \times B$? Note that the Axiom of Comprehension does not apply to the definition appearing in (15). However, the existence of $A \times B$ can be proven:

THEOREM 3.5. *For all sets A and B, there is a set whose elements are precisely all the ordered pairs $\langle a, b \rangle$ in which $a \in A$ and $b \in B$.*

Proof. For any $a \in A$ and $b \in B$, $\{a, b\} \subseteq A \cup B$, so $\{a, b\} \in \mathcal{P}(A \cup B)$. Similarly, $\{a\} \in \mathcal{P}(A \cup B)$, so $\langle a, b \rangle = \{\{a, b\}, \{a\}\} \in \mathcal{P}(\mathcal{P}(A \cup B))$. Thus the definition of $A \times B$ in (15) is equivalent to the following definition, which, while harder to understand than (15), has the virtue that the existence of the set in question is guaranteed by the Axiom of Comprehension (together with the Axioms of Pairing, Unions, and Power Sets):

(16) $A \times B = \{x \in \mathcal{P}(\mathcal{P}(A \cup B)) : \exists a \exists b (a \in A \wedge b \in B \wedge$

      $x = \langle a, b \rangle)\}.$ ■

DEFINITION 3.6.   A set $R \subseteq A \times B$ is called a *relation from A to B*. It is called a *function from A to B* if for each $a \in A$ there is exactly one $b \in B$ such that $\langle a, b \rangle \in R$. It is called a *one-to-one correspondence between A and B* if it is a function from A to B and, in addition, for each $b \in B$ there is exactly one $a \in A$ such that $\langle a, b \rangle \in R$. Two sets A and B are *equinumerous* if there is a one-to-one correspondence between them.

Often, the letter f is used to stand for a function. We write $f : A \to B$ to indicate that f is a function from A to B. If $f : A \to B$, then we say that A is the *domain* of f, and for each $a \in A$ we write $f(a)$ for the unique $b \in B$ such that $\langle a, b \rangle \in f$. For example, if $A = \{p, q\}$, $B = \{x, y, z\}$, and $f = \{\langle p, z \rangle, \langle q, x \rangle\}$, then $f : A \to B$, $f(p) = z$, and $f(q) = x$; however, f is not a one-to-one correspondence between A and B, because no element of A is paired with y in an ordered pair in f. Often, a function from A to B is specified by giving a rule that can be used to determine $f(a)$, for any $a \in A$. From such a rule, the set f can be determined by using the fact that $f = \{\langle a, b \rangle \in A \times B : b = f(a)\}$. For example, if $\mathbb{N}$ is the set of all natural numbers, then we might define a function $f : \mathbb{N} \to \mathbb{N}$ by the rule that, for every natural number n, $f(n) = 2n$. According to the conventions introduced in this paragraph, this means that $f = \{\langle 0, 0 \rangle, \langle 1, 2 \rangle, \langle 2, 4 \rangle, \langle 3, 6 \rangle, \ldots\}$.

Imitating Frege, our next step should be to define the number of elements in a set A to be the extension of the concept *equinumerous with A*, or in other words the set of all sets equinumerous with A. For example, since $\{\varnothing\}$ has one element, the number 1 should be the set:

(17) $\{x : x$ is equinumerous with $\{\varnothing\}\}.$

But, as with our definition of $A \times B$ in (15), we must first check to see if the existence of this set can be proven in ZFC. In the case of $A \times B$ we were able to justify (15) by proving its equivalence to (16) and then appealing to the Axiom of Comprehension. Unfortunately, the definition in (17) fails to define a set.

THEOREM 3.7.   *There is no set containing all the sets equinumerous with $\{\varnothing\}$. In other words, the class of all sets equinumerous with $\{\varnothing\}$ is a proper class.*

Proof. Suppose A is a set containing all sets equinumerous with $\{\varnothing\}$. Then for every set x, $\{x\} \in A$, so $x \in \cup A$. In other words, $\cup A$ contains all sets, contradicting Theorem 3.3.   ∎

Fortunately, this problem is not difficult to solve. Instead of defining the number of elements in a set A to be the collection of all sets equinumerous with A, we will define it to be a particular set chosen from this collection. Thus, the number 0 will be some set with no elements, the number 1 will be some set with one element, and so on. In the case of the number 0, this leaves us no choice: there is only one set with no elements, namely $\varnothing$, so we define the number 0 to be $\varnothing$. But which one-element set should we choose to be the number 1? The choice most commonly used was first suggested by the Hungarian mathematician John von Neumann (1903–57), although the motivation for the choice is already present in Frege's work. Recall that Frege's method of proving that every natural number n has a successor was to show that the number of elements in the set $\{0, 1, 2, \ldots, n\}$ is the successor of n. Why not define the successor of n to be the set $\{0, 1, 2, \ldots, n\}$? This is a recursive definition, which we can use to generate the natural numbers in order as follows:

(18) $0 = \varnothing$,
       $1 = \{0\} = \{\varnothing\}$,
       $2 = \{0, 1\} = \{\varnothing, \{\varnothing\}\}$,
       $3 = \{0, 1, 2\} = \{\varnothing, \{\varnothing\}, \{\varnothing, \{\varnothing\}\}\}$,
       and so on.

Note that if n' is the successor of n, then $n = \{0, 1, 2, \ldots, n-1\}$ and $n' = \{0, 1, 2, \ldots, n-1, n\}$, so $n' = n \cup \{n\}$. This gives us a simple general definition of the successor operation. Combining this with Frege's definition of the ancestral allows us to define the natural numbers:

DEFINITION 3.8.   For any set x, we define the *successor* of x to be the set $S(x) = x \cup \{x\}$. A set F is said to be *hereditary with respect to successor* if $\forall x(x \in F \rightarrow S(x) \in F)$. A set x is a *natural number* if, for every set F, if $\varnothing \in F$ and F is hereditary with respect to successor then $x \in F$.

The existence of the successor of any set x is guaranteed by the Axioms of Pairing and Union. Using this fact repeatedly, we can prove the existence of each of the numbers listed in (18). But what about the set of all natural numbers? It turns out that the existence of this set cannot be proven using the axioms we have introduced so far. In fact, the unprovability of the existence of the set of natural numbers can be proven! Roughly, the idea of the proof is that if we had stopped our construction of the universe of sets at V_ω, then all of the axioms stated so far would be true of the resulting universe of sets, but the universe would not contain any infinite sets, and therefore would not contain the set of all natural numbers. More formally: V_ω is a *model* of all of the axioms stated so far in which there is no set of all natural numbers. Thus, the existence of the set of natural numbers is not implied by the axioms stated so far. We need another axiom:

AXIOM OF INFINITY. There is a set, denoted $\mathbb{N}$, whose elements are precisely all the natural numbers.

Since our definition of the natural numbers is so similar to Frege's, it is not surprising that the Peano Postulates can now be proven from our definition:

THEOREM 3.9.
 (i)   0 *is a natural number.*
 (ii)   *For every natural number x, S(x) is a natural number.*
 (iii)   0 *is not the successor of any natural number.*
 (iv)   *For all natural numbers x and y, if* $x \neq y$ *then* $S(x) \neq S(y)$.
 (v)   *For every set F, if* $0 \in F$ *and F is hereditary with respect to successor then F contains all natural numbers.*

Proof. See exercise 1. ■

Using Peano's Postulates, we can now go on to develop the rest of the theory of the natural numbers. Although we will not carry this project very far here, we will do enough to give the reader some idea of how it is done. We begin by defining addition of natural numbers.

DEFINITION 3.10.   For any $m \in \mathbb{N}$,
 (i) $m + 0 = m$, and
 (ii)   for any $n \in \mathbb{N}, m + S(n) = S(m + n)$.

It may appear at first that Definition 3.10 is circular, since the formula for $m + S(n)$ given in clause (ii) involves addition, but it is actually a recursive definition that can be used to determine the values of $m + 0, m + 1, m + 2$, and so on, in order. For example, here is how the definition can be used to compute $3 + 2$:

(19)   (a) $3 + 0 = 3$ (by 3.10(i)).

     (b) $3 + 1 = 3 + S(0)$

$$= S(3 + 0) \quad \text{(by 3.10(ii))}$$

$$= S(3) \quad \text{(by (a))}$$

$$= 4.$$

    (c) $3 + 2 = 3 + S(1)$

$$= S(3 + 1) \quad \text{(by 3.10(ii))}$$

$$= S(4) \quad \text{(by (b))}$$

$$= 5.$$

Similar reasoning can be used to compute the answer to any addition problem using Definition 3.10. In fact, it can be proven that there is a unique operation on natural numbers that fits the description of addition given in Definition 3.10. More precisely, it can be proven, in ZFC, that there is a unique function $P : \mathbb{N} \times \mathbb{N} \rightarrow \mathbb{N}$ such that, if we define $m + n$ to mean $P(\langle m, n \rangle)$, then equations (i) and (ii) of Definition 3.10 are satisfied (see exercise 2).

All of the familiar properties of addition of natural numbers can be proven from Definition 3.10. Since the definition is recursive, it should not be surprising that many of these proofs use the method of mathematical induction. For example, here is the proof that addition of natural numbers is associative:

THEOREM 3.11.   *For all natural numbers m, n, and k, $m + (n + k) = (m + n) + k$.*

Proof. Let m and n be any two natural numbers. We now prove by induction on k that $\forall k (m + (n + k) = (m + n) + k)$.

Base case: If $k = 0$, then

$$m + (n + k) = m + (n + 0)$$

$$= m + n \quad \text{(by 3.10(i))}$$

$$= (m + n) + 0 \quad \text{(by 3.10(i))}$$

$$= (m + n) + k.$$

Induction step: Let k be an arbitrary natural number, and assume that $m + (n + k) = (m + n) + k$. Then

$$
\begin{aligned}
m + (n + S(k)) &= m + S(n + k) &&\text{(by 3.10(ii))} \\
&= S(m + (n + k)) &&\text{(by 3.10(ii))} \\
&= S((m + n) + k) &&\text{(by the inductive hypothesis)} \\
&= (m + n) + S(k) &&\text{(by 3.10(ii)).} \qquad \blacksquare
\end{aligned}
$$

It is perhaps worthwhile to point out why the use of mathematical induction in this proof is justified by clause (5) of Theorem 3.9. If we let $F = \{k \in \mathbb{N} : m + (n + k) = (m + n) + k\}$, then the base case of the induction proof shows that $0 \in F$, and the induction step shows that F is hereditary with respect to successor. Thus, by Theorem 3.9(5), F contains all natural numbers, so $\forall k(m + (n + k) = (m + n) + k)$.

Similar proofs can be given for other familiar laws of addition, such as the commutative law $(m + n = n + m)$ and the cancellation law (if $m + k = n + k$, then $m = n$) (see exercises 3–5). And similar definitions can be given for other operations on natural numbers. For example, multiplication is defined as follows:

DEFINITION 3.12.   For any $m \in \mathbb{N}$,
  (i) $m \cdot 0 = 0$, and
  (ii)   for any $n \in \mathbb{N}$, $m \cdot S(n) = (m \cdot n) + m$.

As an example of how this definition can be used to derive the basic properties of multiplication of natural numbers, we prove the distributive law:

THEOREM 3.13.   *For all natural numbers m, n, and k, $m \cdot (n + k) = (m \cdot n) + (m \cdot k)$.*

Proof. Let m and n be any two natural numbers. We now prove by induction on k that $\forall k(m \cdot (n + k) = (m \cdot n) + (m \cdot k))$.

Base case: If $k = 0$, then

$$
\begin{aligned}
m \cdot (n + k) &= m \cdot (n + 0) \\
&= m \cdot n &&\text{(by 3.10(i))} \\
&= (m \cdot n) + 0 &&\text{(by 3.10(i))} \\
&= (m \cdot n) + (m \cdot 0) &&\text{(by 3.12(i))} \\
&= (m \cdot n) + (m \cdot k).
\end{aligned}
$$

Induction step: Let k be any natural number, and assume that $m \cdot (n + k) = (m \cdot n) + (m \cdot k)$. Then

$$
\begin{aligned}
m \cdot (n + S(k)) &= m \cdot S(n + k) && \text{(by 3.10(ii))} \\
&= (m \cdot (n + k)) + m && \text{(by 3.12(ii))} \\
&= ((m \cdot n) + (m \cdot k)) + m && \text{(by the inductive hypothesis)} \\
&= (m \cdot n) + ((m \cdot k) + m) && \text{(by 3.11)} \\
&= (m \cdot n) + (m \cdot S(k)) && \text{(by 3.12(ii)).} \qquad \blacksquare
\end{aligned}
$$

As a final illustration of the development of the theory of the natural numbers, we define the ordering of the natural numbers:

DEFINITION 3.14.   For any natural numbers m and n, we say that m is *less than* n, and write $m < n$, if there is a natural number $k \neq 0$ such that $m + k = n$.

Using this definition, we can prove all of the familiar properties of the ordering of the natural numbers, such as the trichotomy law: $\forall m \forall n (m < n \lor m = n \lor n < m)$ (see exercise 8).

The natural numbers are used primarily for counting how many elements are in a set. If n is a natural number, then we say that a set A *has n elements* if A is equinumerous with n. A set A is called *finite* if it has n elements for some natural number n, and otherwise it is called *infinite*.

Of course, there are applications of numbers for which the natural numbers are not adequate; sometimes we need other numbers, such as negative numbers and fractions. Thus, we will need to define number systems containing these other kinds of numbers. We will define three more number systems: the integers, the rational numbers, and the real numbers. Before we give the formal definitions, it will be helpful to say informally what these number systems are.

The set of all integers, denoted $\mathbb{Z}$, contains all positive and negative whole numbers:

(20) $\mathbb{Z} = \{\ldots, -3, -2, -1, 0, 1, 2, 3, \ldots\}$.

The set of all rational numbers, denoted $\mathbb{Q}$, contains all numbers that can be expressed as fractions. This includes all integers, since an integer can always be written as a fraction whose denominator is 1 (for example, $17 = 17/1$):

(21) $\mathbb{Q} = \{p/q : p \text{ and } q \text{ are integers}, q \neq 0\} = \{1/2, -5/3, 17, \ldots\}$.

The set of all real numbers, denoted $\mathbb{R}$, can be thought of as representing all points on a number line. It includes all rational numbers, but it also includes some numbers, such as $\sqrt{2}$ and π, that cannot be written as fractions.

We are now ready to turn to the formal definition of the integers, and the derivation of their basic properties from the ZFC axioms. Our definition will be motivated by thinking of the integers as answers to subtraction problems involving pairs of natural numbers. For example, -3 is the answer to the subtraction problem $4 - 7$, which we can think of as being determined by the pair of natural numbers $\langle 4, 7 \rangle$. Note that it is important to use *ordered* pairs here; the pair $\langle 7, 4 \rangle$ determines the subtraction problem $7 - 4$, whose answer is 3, not -3. This suggests that we could simply define integers to be ordered pairs of natural numbers. For example, the integer -3 would be defined to be the pair $\langle 4, 7 \rangle$.

Unfortunately, this approach will not work, because different subtraction problems sometimes have the same answer. For example, $8 - 11$ is another problem whose answer is -3. Which ordered pair will the integer -3 be, $\langle 4, 7 \rangle$ or $\langle 8, 11 \rangle$? Our solution to this difficulty will involve grouping together those ordered pairs of natural numbers that represent subtraction problems with the same answer. For example, according to the definition we will eventually give, the integer -3 will be the set $\{\langle 4, 7 \rangle, \langle 8, 11 \rangle, \langle 2, 5 \rangle, \ldots\}$, whose elements are all ordered pairs of natural numbers that determine subtraction problems with answer -3.

The idea of grouping together objects that share some property is a common one in mathematics. In fact, it is the same method that Frege used in his definition of the cardinal numbers, which involved grouping together concepts that are equinumerous. Before explaining how this method can be used to define the integers, it would be worthwhile to discuss more generally how such grouping is usually accomplished. As was mentioned in Chapter 2, the technique used most often in mathematics involves the use of equivalence relations. We begin our discussion with the idea of a relation on a set.

DEFINITION 3.15.   Suppose A is a set. A set $R \subseteq A \times A$ is called a *relation on* A. In other words, in the language of Definition 3.6, a relation on A is a relation from A to A. If $x, y \in A$, we sometimes write xRy to mean $\langle x, y \rangle \in R$.

For example, let A be the set of all words in the English language, let $R = \{\langle x, y \rangle \in A \times A : x \text{ and } y \text{ begin with the same letter}\}$, and let $S = \{\langle x, y \rangle \in A \times A : x \text{ and } y \text{ have a letter in common}\}$. Then R and S are both relations on A. Using the shorthand notation introduced in Definition 3.15 we could write "appleRapricot", since "apple" and "apricot" both start with "a", and therefore $\langle \text{apple}, \text{apricot} \rangle \in R$.

Similarly, since the words "apple" and "berry" both contain the letter "e" but start with different letters, we can write "apple$S$berry", but not "apple$R$berry".

An interesting property of the relation R is that it groups the elements of A into 26 clusters: words that begin with "a", words that begin with "b", ..., words that begin with "z". Ordered pairs consisting of words from the same cluster are elements of R, while pairs of words from different clusters are not. These clusters form what is called a *partition* of A: a collection of nonempty subsets of A with the property that every element of A belongs to exactly one of these subsets. The relation S, on the other hand, does not determine a partition of A in this way. What is responsible for this difference?

Notice that the relation R has the following properties. Since every word x starts with the same letter as itself, we always have xRx. If xRy for some words x and y, then x and y start with the same letter, so yRx. And if xRy and yRz, then x, y, and z must all start with the same letter, so we must also have xRz. The relation S has the first two of these properties, but not the third. It is useful to have names for these three properties:

DEFINITION 3.16.   Suppose R is a relation on a set A.
   (i) R is called *reflexive* if, for all x in A, xRx.
   (ii) R is called *symmetric* if, for all x and y in A, if xRy then yRx.
   (iii) R is called *transitive* if, for all x, y, and z in A, if xRy and yRz then xRz.
   (iv) R is called an *equivalence relation on A* if it is reflexive, symmetric, and transitive.

Thus, the "same first letter" relation is an equivalence relation on the set of English words. In fact, it is not hard to see that any relation on any set that determines a partition of that set into clusters in the way described earlier will be an equivalence relation. And it can be shown that the converse is true as well: any equivalence relation on any set will determine a partition of the set into clusters (see exercise 9).

If R is an equivalence relation on A and $x \in A$, then the set of elements of A that are paired with x in the relation R is called the *equivalence class of x with respect to R*, and is denoted $[x]_R$, or just $[x]$ if R is clear from the context. In other words,

(22) $[x]_R = \{y \in A : xRy\}$.

For example, for the "same first letter" relation R discussed earlier, $[\text{apple}]_R$ would be the set of all words that begin with "a". In general,

$[x]_R$ is the set of elements of A that are in the same cluster as x. The set whose elements are all of these equivalence classes is called A *mod* R, and is denoted A/R. Thus,

(23) $\quad A/R = \{[x]_R : x \in A\} = \{X \in \mathcal{P}(A) : \text{for some } x \in A,\ X = [x]_R\}.$

For example, in the "same first letter" example, A/R is a set with 26 elements. Each of those elements is itself a set containing all words beginning with a particular letter. In other words, $A/R = \{\{\text{apple},$ apricot, $\dots\}, \{\text{berry, bubble}, \dots\}, \dots, \{\text{zebra, zoologist}, \dots\}\}$.

With this background, we can now return to the formal definition of the integers. Recall that our plan is to think of integers as the answers to subtraction problems, and to think of subtraction problems as being represented by ordered pairs of natural numbers. Thus, we begin by forming the set $\mathbb{N} \times \mathbb{N}$ of all ordered pairs of natural numbers. In order to group together pairs of natural numbers that represent subtraction problems with the same answer, we now define an equivalence relation on $\mathbb{N} \times \mathbb{N}$. Often in mathematics equivalence relations are represented not by letters but by symbols. We will call our equivalence relation $\sim$; in other words, we will write $\langle a, b \rangle \sim \langle c, d \rangle$ to indicate that the pairs $\langle a, b \rangle$ and $\langle c, d \rangle$ should be grouped together.

It is tempting to define $\langle a, b \rangle \sim \langle c, d \rangle$ to mean that $a - b = c - d$, but this is unacceptable because we have not yet defined subtraction. Fortunately, the equation $a - b = c - d$ can be rewritten as $a + d = b + c$, an equation involving only addition of natural numbers, which we have already defined. Thus, we define the relation $\sim$ as follows:

DEFINITION 3.17. For any natural numbers $a, b, c,$ and d, $\langle a, b \rangle \sim \langle c, d \rangle$ means $a + d = b + c$.

THEOREM 3.18. *The relation $\sim$ is an equivalence relation on $\mathbb{N} \times \mathbb{N}$.*

Proof. We will leave the proofs that $\sim$ is reflexive and symmetric as exercises for the reader (see exercise 10). To prove transitivity, suppose that $\langle a, b \rangle \sim \langle c, d \rangle$ and $\langle c, d \rangle \sim \langle e, f \rangle$. Then, by Definition 3.17, $a + d = b + c$ and $c + f = d + e$. Adding these equations gives us $a + d + c + f = b + c + d + e$ or, by associativity and commutativity of addition, $(a + f) + (c + d) = (b + e) + (c + d)$. By the cancellation law for addition, $a + f = b + e$, so $\langle a, b \rangle \sim \langle e, f \rangle$. ∎

Since $\sim$ is an equivalence relation on $\mathbb{N} \times \mathbb{N}$, it partitions $\mathbb{N} \times \mathbb{N}$ into clusters. The cluster containing an ordered pair $\langle a, b \rangle$ is the equivalence class of $\langle a, b \rangle$, and is denoted $[\langle a, b \rangle]$. We will define these equivalence classes to be the integers. For example, $[\langle 4, 7 \rangle] = \{\langle c, d \rangle \in \mathbb{N} \times \mathbb{N} : \langle 4, 7 \rangle \sim \langle c, d \rangle\} = \{\langle c, d \rangle \in \mathbb{N} \times \mathbb{N} : d + 4 = c + 7\} = \{\langle 4, 7 \rangle, \langle 8, 11 \rangle, \langle 2, 5 \rangle, \ldots\}$. This is the set that we said earlier would turn out to be the integer -3.

DEFINITION 3.19.   The equivalence classes $[\langle a, b \rangle]$, for $a, b \in \mathbb{N}$, are called *integers*. The set of all integers is denoted $\mathbb{Z}$. In other words, $\mathbb{Z} = (\mathbb{N} \times \mathbb{N})/\sim$.

Suppose that x and y are integers. How should we define $x + y$? Since, by Definition 3.19, x and y are sets of pairs of natural numbers, the following procedure seems reasonable. Choose pairs of natural numbers $\langle a, b \rangle \in x$ and $\langle c, d \rangle \in y$. Intuitively, this means that x is the answer to the subtraction problem $a - b$, and y is $c - d$. But then $x + y$ should be $(a - b) + (c - d) = (a + c) - (b + d)$, which is the answer to another subtraction problem. This suggests that we should define $x + y$ to be the integer $[\langle a + c, b + d \rangle]$.

There is one potential problem with this definition. The first step in our procedure for adding x and y was to *choose* pairs $\langle a, b \rangle \in x$ and $\langle c, d \rangle \in y$. This suggests that our definition of $x + y$ might be ambiguous, since the answer might depend on which pairs we choose. What if we had made different choices – say, $\langle a', b' \rangle \in x$ and $\langle c', d' \rangle \in y$? Then we would have computed $x + y$ to be $[\langle a' + c', b' + d' \rangle]$. Is this the same as our original answer, $[\langle a + c, b + d \rangle]$? Fortunately, it is, but this requires proof. To begin the proof, first note that, since $\langle a, b \rangle$ and $\langle a', b' \rangle$ come from the same cluster x, we must have $\langle a, b \rangle \sim \langle a', b' \rangle$, and similarly $\langle c, d \rangle \sim \langle c', d' \rangle$. And to prove that $[\langle a + c, \ b + d \rangle] = [\langle a' + c', b' + d' \rangle]$, it suffices to show that $\langle a + c, b + d \rangle \sim \langle a' + c', b' + d' \rangle$. Thus, the following theorem shows that our definition of addition is unambiguous:

THEOREM 3.20.   *Suppose* $\langle a, b \rangle \sim \langle a', b' \rangle$ *and* $\langle c, d \rangle \sim \langle c', d' \rangle$. *Then* $\langle a + c, b + d \rangle \sim \langle a' + c', b' + d' \rangle$.

Proof. Since $\langle a, b \rangle \sim \langle a', b' \rangle$ and $\langle c, d \rangle \sim \langle c', d' \rangle$, by Definition 3.17 we have $a + b' = b + a'$ and $c + d' = d + c'$. Adding these equations gives us $a + c + b' + d' = b + d + a' + c'$, so $\langle a + c, b + d \rangle \sim \langle a' + c', b' + d' \rangle$. ∎

Note that since $\langle a, b \rangle \in x$, we have $x = [\langle a, b \rangle]$, and similarly $y = [\langle c, d \rangle]$. Thus we can rewrite our definition of addition as follows:

DEFINITION 3.21.   For any natural numbers a, b, c, and d,

(24) $[\langle a, b \rangle] + [\langle c, d \rangle] = [\langle a + c, b + d \rangle]$.

As an example of the use of this definition, we can compute $[\langle 4, 7 \rangle] + [\langle 3, 2 \rangle] = [\langle 7, 9 \rangle]$. In more ordinary notation, this equation says that $-3 + 1 = -2$.

It might seem at first that Definition 3.21 is circular, since addition is being used in the definition of addition. But this is not the case, because the "+" on the left side of (24) stands for addition of integers, but the "+" on the right side stands for addition of natural numbers. Thus addition of integers is being defined in terms of addition of natural numbers, which has already been defined. As a result, many of the basic properties of addition of integers follow immediately from the corresponding properties of addition of natural numbers. For example:

THEOREM 3.22.   *Addition of integers is commutative; that is, for all integers x and y, $x + y = y + x$.*

Proof. Suppose $x = [\langle a, b \rangle]$ and $y = [\langle c, d \rangle]$. Then

$$
\begin{aligned}
x + y &= [\langle a, b \rangle] + [\langle c, d \rangle] \\
&= [\langle a + c, b + d \rangle] && \text{(by Definition 3.21)} \\
&= [\langle c + a, d + b \rangle] && \text{(by commutativity of + for natural} \\
& && \text{numbers)} \\
&= [\langle c, d \rangle] + [\langle a, b \rangle] && \text{(by Definition 3.21)} \\
&= y + x.
\end{aligned}
$$
∎

The definition of multiplication of integers is handled in a similar way. If $x = [\langle a, b \rangle]$ and $y = [\langle c, d \rangle]$, then intuitively we think of x and y as the answers to the subtraction problems $a - b$ and $c - d$. Thus $x \cdot y$ should be $(a - b) \cdot (c - d) = ac - ad - bc + bd = (ac + bd) - (ad + bc)$. We therefore define multiplication of integers as follows:

DEFINITION 3.23.   For any natural numbers a, b, c, and d,

(25) $[\langle a, b \rangle] \cdot [\langle c, d \rangle] = [\langle ac + bd, ad + bc \rangle]$.

As with addition, we need a theorem to show that this definition is unambiguous. We leave the proof of this theorem as an exercise for the reader (see exercise 11).

THEOREM 3.24.   *Suppose* $\langle a, b \rangle \sim \langle a', b' \rangle$ *and* $\langle c, d \rangle \sim \langle c', d' \rangle$. *Then* $\langle ac + bd, ad + bc \rangle \sim \langle a'c' + b'd', a'd' + b'c' \rangle$.

Since one of our purposes in defining the integers was to provide answers to subtraction problems, we should also point out that we can now define subtraction of integers. The definition is motivated by the algebraic equation $(a - b) - (c - d) = (a + d) - (b + c)$.

DEFINITION 3.25.   For any natural numbers a, b, c, and d,

(26) $[\langle a, b \rangle] - [\langle c, d \rangle] = [\langle a + d, b + c \rangle]$.

Once again, a theorem must be proven to show that this definition is unambiguous (see exercise 12).

Finally, we can also distinguish between positive and negative integers, and use this distinction to define the ordering of the integers. Note that the "<" used in (27) below refers to the ordering of the natural numbers, which has already been defined, so this definition is not circular.

DEFINITION 3.26.   The set of *positive* integers is the set

(27) $\mathbb{Z}^+ = \{[\langle a, b \rangle] : b < a\}$.

For any integers x and y, we define $x < y$ to mean that $y - x$ is positive.

There is one aspect of our definition of the integers that may be troubling the reader at this point. We ordinarily think of the natural numbers as forming a subset of the integers, but as we have defined them this is not true. For example, the natural number 2 is not an integer, although there is an integer that we think of, intuitively, as the answer to the subtraction problem $2 - 0$, namely the integer $[\langle 2, 0 \rangle]$. Similarly, we might think of any natural number a as corresponding to the integer $[\langle a, 0 \rangle]$, but the natural number and the integer are not the same. All of the properties and operations we have defined on natural numbers match up with the corresponding properties and operations on the corresponding integers. For example, according to the definition of the addition of natural numbers, $2 + 3 = 5$, and according to the definition of addition of integers, $[\langle 2, 0 \rangle] + [\langle 3, 0 \rangle] = [\langle 5, 0 \rangle]$. In more technical language, we can say that there is a subset of $\mathbb{Z}$ that is *isomorphic* to $\mathbb{N}$, but $\mathbb{N}$ itself is not a subset of $\mathbb{Z}$.

Mathematicians usually say that this problem is to be dealt with by "identifying" the natural number a with the integer $[\langle a, 0 \rangle]$. On the face of it this suggestion seems absurd; if two mathematical objects are not identical, how can it be correct to say that they are? Perhaps a better interpretation is to say that mathematicians choose to ignore the difference between a and $[\langle a, 0 \rangle]$. This choice reflects the tendency of mathematicians to regard the differences between isomorphic mathematical structures as unimportant. The set $\mathbb{N}$ and the set of integers of the form $[\langle a, 0 \rangle]$ are isomorphic, so the differences between them are mathematically unimportant.

Another possibility would be to declare that the elements of $\mathbb{N}$ are not really the natural numbers, but are rather "proto-natural numbers." They were defined just so that we could use them to define the integers, and can now be ignored. The "real" natural numbers are the integers of the form $[\langle a, 0 \rangle]$. This would resolve the difficulty by making the natural numbers a subset of the integers, although if we take this approach then we will soon find that our integers will have to be renamed "proto-integers" when we move on to define the rational numbers.

Just as the integers provided the answers to subtraction problems involving natural numbers, we can think of the rational numbers as providing answers to division problems involving integers. Thus, to define the rational numbers, we will think of an ordered pair of integers $\langle x, y \rangle$ as representing the fraction x/y, and we will group together the pairs $\langle x, y \rangle$ and $\langle u, v \rangle$ if $x/y = u/v$ or, equivalently, $xv = yu$. Of course, the denominator of a fraction must not be zero, so we only consider pairs of integers in which the second integer is nonzero. Since the definition of the rational numbers is so similar to the definition of the integers, we will skip most of the details and just briefly list the main points.

DEFINITION 3.27. Let $\mathcal{F} = \{\langle x, y \rangle : x, y \in \mathbb{Z}, y \neq 0\}$. For $\langle x, y \rangle$, $\langle u, v \rangle \in \mathcal{F}$, we define $\langle x, y \rangle \equiv \langle u, v \rangle$ to mean that $xv = yu$.

THEOREM 3.28. *The relation $\equiv$ is an equivalence relation on $\mathcal{F}$.*

Proof. See exercise 15. ■

DEFINITION 3.29. The equivalence classes $[\langle x, y \rangle]$ of pairs $\langle x, y \rangle \in \mathcal{F}$ are called *rational numbers*. The set of all rational numbers is denoted $\mathbb{Q}$. In other words, $\mathbb{Q} = \mathcal{F} / \equiv$.

For example, the rational number $1/2$ is the set $[\langle 1, 2 \rangle] = \{\langle 1, 2 \rangle, \langle 2, 4 \rangle, \langle 3, 6 \rangle, \ldots\}$, where the numerals in this equation stand for integers. The rational number 3 is the set $[\langle 3, 1 \rangle] = \{\langle 3, 1 \rangle, \langle 6, 2 \rangle, \langle 9, 3 \rangle, \ldots\}$. Note that the rational number 3 is different from the integer 3. Imitating our previous definition, we deal with this situation by identifying each integer x with the corresponding rational number $[\langle x, 1\rangle]$.

The familiar grade school formulas for performing arithmetical operations on fractions motivate the next definitions. As before, for each definition one must prove a theorem showing that the definition is unambiguous (see exercise 16).

DEFINITION 3.30.   For any $\langle x, y\rangle, \langle u, v\rangle \in \mathcal{F}$,
   (i) $[\langle x, y\rangle] + [\langle u, v\rangle] = [\langle xv + yu, yv\rangle]$.
   (ii) $[\langle x, y\rangle] - [\langle u, v\rangle] = [\langle xv - yu, yv\rangle]$.
   (iii) $[\langle x, y\rangle] \cdot [\langle u, v\rangle] = [\langle xu, yv\rangle]$.
   (iv) If $u \neq 0$, then $[\langle x, y\rangle] \div [\langle u, v\rangle] = [\langle xv, yu\rangle]$.
   (v)   The set of *positive* rational numbers is the set $\mathbb{Q}^+ = \{[\langle x, y\rangle]$: either $x > 0$ and $y > 0$, or $x < 0$ and $y < 0\}$. For rational numbers p and q, we define $p < q$ to mean that $q - p$ is positive.

For example, by 3.30(i) we have $[\langle 1, 2\rangle] + [\langle 2, 3\rangle] = [\langle 1 \cdot 3 + 2 \cdot 2,$ $2 \cdot 3\rangle] = [\langle 7, 6\rangle]$. In more familiar notation, this says that $1/2 + 2/3 = 7/6$. It is straightforward, but tedious, to verify that all of the familiar rules of algebra for rational numbers can be proven from these definitions.

Finally, we turn now to the definition of the real numbers. We introduced the integers to provide answers to subtraction problems, and the rational numbers to provide answers to division problems. The problem the real numbers will solve is more subtle. As we saw in Theorem 1.1, $\sqrt{2}$ is irrational, so one reason for introducing the real numbers is to provide answers to square root problems. But many other irrational numbers, such as π, are not square roots of any rational numbers. What, precisely, is the shortcoming of the rational numbers that will be remedied by introducing the real numbers?

One way to answer this question is to imagine that all of the rational numbers have been marked off on a number line. If the line includes points at all of the rational numbers, but no others, then the line will be full of holes. For example, the squares of some positive rational numbers are less than 2, and some are greater than 2, but since $\sqrt{2}$ is irrational, there is no point on the rational number line at the boundary between these two sets of rational numbers. This boundary point is therefore a hole in the rational number line. Similarly, although there are rational numbers very close to π, no fraction is exactly equal to π, so there is a hole in the rational number line where π belongs. The purpose of defining the real numbers is to fill in the holes in the rational number line, producing a continuous number line with no gaps. The property of the real number line that guarantees that it has no holes in it is called *completeness* (see Definition 3.40 and Theorem 3.41 below). The completeness of the real numbers is crucial for making the proofs of many of the theorems in calculus work.

How does one specify the location of a hole in the rational number line? We can get an idea for how to do this by examining the familiar decimal notation for real numbers. For example, consider the decimal expansions of $\sqrt{2}$ and π:

(28) $\quad \sqrt{2} = 1.41421356\ldots,$

$\quad\quad \pi = 3.14159265\ldots.$

The "..." at the end of each of these decimal expansions indicates that the list of digits goes on forever. In fact, no irrational number can be represented by a finite decimal expansion, because such decimal expansions always represent rational numbers. For example, 1.414 is equal to 1414/1000, a rational number, so, although it is a fairly good approximation to $\sqrt{2}$, it cannot be exactly equal to $\sqrt{2}$, which is irrational. We can get a better approximation by adding more digits – for example, $1.4142 = 14142/10000$ is a better approximation to $\sqrt{2}$ than 1.414 – but no finite decimal expansion can give the value of $\sqrt{2}$ exactly. Since the lists of digits in (28) are infinitely long, we can continue this process of producing better and better rational approximations forever.

This suggests that we could specify the location of a hole in the rational number line by giving an infinite sequence of rational numbers getting closer and closer to the hole. For example, $\sqrt{2}$ and π could be picked out by the following infinite sequences:

(29) $\quad \sqrt{2} : 1, 1.4, 1.41, 1.414, 1.4142, 1.41421, \ldots$

$\quad\quad \pi : 3, 3.1, 3.14, 3.141, 3.1415, 3.14159, \ldots$

Note that each number in each sequence is rational, since it is given by a finite decimal expansion.

Before proceeding further, we should say how we can talk about infinite sequences of rational numbers in ZFC. Since the language of set theory allows us to refer only to sets, this means we must represent infinite sequences as sets. We will define the infinite sequence $a_1, a_2, a_3,$ ...to be the set $\{\langle 1, a_1 \rangle, \langle 2, a_2 \rangle, \langle 3, a_3 \rangle, \ldots\}$. Note that, according to Definition 3.6, this set is a function with domain $\mathbb{Z}^+$. Thus, we can state the definition of sequence more precisely as follows:

DEFINITION 3.31. Suppose A is a set. Then an *infinite sequence* of elements of A is a function $S : \mathbb{Z}^+ \to A$. For each $n \in \mathbb{Z}^+$, $S(n)$ is called the *nth term of the sequence*.

Following the notation common in calculus, we will write $\{a_n\}_{n=1}^{\infty}$, or sometimes just $\{a_n\}$, to represent the sequence whose nth term is a_n. In other words, $\{a_n\}_{n=1}^{\infty} = \{\langle 1, a_1 \rangle, \langle 2, a_2 \rangle, \langle 3, a_3 \rangle, \ldots\}$. More informally, we will sometimes represent the sequence $\{\langle 1, a_1 \rangle, \langle 2, a_2 \rangle, \langle 3, a_1 \rangle, \ldots\}$ as a list: $a_1, a_2, a_3, \ldots$.

Motivated by the examples in (29), one might be tempted now to define a real number to be an infinite sequence of rational numbers, but there are two problems with this proposal. First, there is more than one infinite sequence of rational numbers that approximates more and more closely the location of any hole in the rational number line. The solution to this problem is easy: we will define an equivalence relation on infinite sequences of rational numbers, grouping together different sequences that approach the same location on the rational number line, and define real numbers to be equivalence classes of sequences.

But there is a second problem that we will also have to deal with. Not every infinite sequence of rational numbers approaches a particular location on the rational number line. For example, consider the following sequences:

(30) $1, 3, 1.4, 3.1, 1.41, 3.14, 1.414, 3.141, \ldots$

       $1, 2, 3, 4, 5, 6, \ldots$

The first sequence cannot seem to make up its mind between $\sqrt{2}$ and π; the terms of the second just get larger and larger, not approaching any particular location on the number line. Neither of these sequences can be used to specify a real number.

Thus, before defining our equivalence relation on infinite sequences of rational numbers, we must first restrict our attention to sequences that approach some location on the number line. Note that we cannot simply say that we will only use those sequences with the property that there is some location on the number line that the sequence approaches. To formulate this property, we would have to say exactly what is meant by the phrase "location on the number line," and that is precisely the concept that we are trying to pin down by using infinite sequences! We must formulate some property of the sequence itself, making no reference to locations on the number line, that will distinguish sequences that approach locations on the number line from sequences that do not.

Consider a sequence of rational numbers $\{a_n\}_{n=1}^{\infty}$. The key to solving our problem is to observe that, if the terms in this sequence get closer and closer to some location on the number line, then they must also get closer and closer to *each other*. How close must they get to each other? Must they get within 1/10 of each other? or 1/100? If the sequence is to converge on a *precise* location on the number line, then surely the answers to these questions must be "yes". In fact, for *every* positive integer k the terms must get within $1/k$ of each other. By gradually refining this statement, we can develop a precise characterization of the sequences we are interested in. The following statements capture more and more precisely the idea we are after:

(31) $\forall k \in \mathbb{Z}^+$(the terms of the sequence eventually get within $1/k$
        of each other).
        $\forall k \in \mathbb{Z}^+$(from some point on, the terms are all within $1/k$
        of each other).
        $\forall k \in \mathbb{Z}^+ \exists N \in \mathbb{Z}^+$(beyond term number N, all terms are within
        $1/k$ of each other).
        $\forall k \in \mathbb{Z}^+ \exists N \in \mathbb{Z}^+ \forall m, n > N(a_m$ and a_n are within $1/k$ of each
        other).

Finally, we use the fact that the distance between a_m and a_n can be written using the formula $|a_m - a_n|$ to define the property we are interested in. Sequences with this property are named after the French mathematician Augustin Cauchy (1789–1857), who first recognized the importance of such sequences.

DEFINITION 3.32.   A sequence of rational numbers $\{a_n\}$ is called a *Cauchy sequence* if

(32) $\forall k \in \mathbb{Z}^+ \exists N \in \mathbb{Z}^+ \forall m, n > N(|a_m - a_n| < 1/k)$.

Since this definition is somewhat complicated, it might be helpful to look at an example. Consider the sequence $\{a_n\}$ whose terms are given by the formula $a_n = 1 + 3/n$. The first few terms of this sequence are $4, 5/2, 2, 7/4, \ldots$. The easiest way to show that this sequence is Cauchy is to specify, for any given positive integer k, the integer N whose existence is required in (32). It turns out that the choice $N = 3k$ works, as the following proof shows.

Proof that $\{a_n\}$ is Cauchy. Let k be an arbitrary positive integer. Let $N = 3k$. Suppose $m, n > N$.

Case 1: $m \leq n$. Then $3/m \geq 3/n$, so

$$|a_m - a_n| = \left|\left(1 + \frac{3}{m}\right) - \left(1 + \frac{3}{n}\right)\right| = \left|\frac{3}{m} - \frac{3}{n}\right| = \frac{3}{m} - \frac{3}{n}$$
$$< \frac{3}{m} < \frac{3}{N} = \frac{3}{3k} = \frac{1}{k}.$$

Case 2: $m > n$. Then $3/m < 3/n$, so

$$|a_m - a_n| = \left|\left(1 + \frac{3}{m}\right) - \left(1 + \frac{3}{n}\right)\right| = \left|\frac{3}{m} - \frac{3}{n}\right| = \frac{3}{n} - \frac{3}{m}$$
$$< \frac{3}{n} < \frac{3}{N} = \frac{3}{3k} = \frac{1}{k}.$$

Since these cases cover all of the possibilities, we can conclude that $\forall m, n > N(|a_m - a_n| < 1/k)$. Since k was an arbitrary positive integer, it follows that $\forall k \in \mathbb{Z}^+ \exists N \in \mathbb{Z}^+ \forall m, n > N(|a_m - a_n| < 1/k)$, as required. ∎

It can be shown that sequences that are generated from decimal expansions, like those in (29), are always Cauchy (see exercise 21). On the other hand, neither of the sequences in (30) is Cauchy. Successive terms in both sequences differ by more than 1/2, so when $k = 2$ there is no N as required in the definition of Cauchy for either sequence.

Let C be the set of all Cauchy sequences of rational numbers. We now define an equivalence relation on C, grouping together sequences that approach the same location on the number line. Once again, we can phrase the definition without having to refer to "locations on the number line" by using the fact that if the terms of two sequences are both close to the same location, then they must also be close to each other.

DEFINITION 3.33.   Suppose $\{a_n\}$ and $\{b_n\}$ are Cauchy sequences of rational numbers. Then $\{a_n\} \approx \{b_n\}$ means

(33) $\forall k \in \mathbb{Z}^+ \exists N \in \mathbb{Z}^+ \forall n > N(|a_n - b_n| < 1/k)$.

THEOREM 3.34.   *The relation $\approx$ is an equivalence relation on C.*

Proof. We will leave the proofs that $\approx$ is reflexive and symmetric as exercises (see exercise 22), but we will give the proof of transitivity because it illustrates several of the tricks that are often used in proofs involving Cauchy sequences. In particular, in one step of the proof we will use the fact that, for any rational numbers a and b, $|a + b| \leq |a| + |b|$. This inequality is known as the *triangle inequality*.

Suppose $\{a_n\} \approx \{b_n\}$ and $\{b_n\} \approx \{c_n\}$. To prove that $\{a_n\} \approx \{c_n\}$, let k be an arbitrary positive integer. Then $2k$ is also a positive integer, so since $\{a_n\} \approx \{b_n\}$, there is a positive integer N_1 such that

$$(34) \quad \forall n > N_1 \left(|a_n - b_n| < \frac{1}{2k} \right).$$

Similarly, since $\{b_n\} \approx \{c_n\}$, there is a positive integer N_2 such that

$$(35) \quad \forall n > N_2 \left(|b_n - c_n| < \frac{1}{2k} \right).$$

Let $N = \max(N_1, N_2) =$ the larger of N_1 and N_2, and note that $N \geq N_1$ and $N \geq N_2$. Now suppose that $n > N$. Then $n > N_1$ and $n > N_2$, so by (34) and (35) we have

$$(36) \quad |a_n - b_n| < \frac{1}{2k} \quad \text{and} \quad |b_n - c_n| < \frac{1}{2k}.$$

Therefore

$$
\begin{aligned}
|a_n - c_n| &= |(a_n - b_n) + (b_n - c_n)| \\
&\leq |a_n - b_n| + |b_n - c_n| \quad &\text{(by the triangle inequality)} \\
&< \frac{1}{2k} + \frac{1}{2k} = \frac{1}{k} \quad &\text{(by (36)).}
\end{aligned}
$$

Thus, $\forall n > N(|a_n - c_n| < 1/k)$. Since k was arbitrary, it follows that $\forall k \in \mathbb{Z}^+ \exists N \in \mathbb{Z}^+ \forall n > N(|a_n - c_n| < 1/k)$, which proves that $\{a_n\} \approx \{c_n\}$. ■

DEFINITION 3.35. The equivalence classes $[\{a_n\}]$ of sequences $\{a_n\} \in C$ are called *real numbers*. The set of all real numbers is denoted $\mathbb{R}$. In other words, $\mathbb{R} = C/\approx.$[6]

---

[6] This definition of the real numbers was first published by Cantor in 1872. Some readers may be familiar with another definition of the real numbers, based on partitions of the rational numbers into two sets such that all numbers in one

For example, $\pi = [\{b_n\}]$, where $\{b_n\}$ is the sequence 3, 3.1, 3.14, 3.141, ... Although the real numbers as we have defined them do not contain the rational numbers as a subset, they contain numbers that we can identify with the rational numbers. For example, the real number 3 is $[\{c_n\}]$, where $\{c_n\}$ is the sequence 3, 3, 3, 3, ... In general, we identify any rational number r with the real number $[\{d_n\}]$, where $\{d_n\}$ is given by the formula $d_n = r$. It is easy to verify that these constant sequences are Cauchy.

The definitions of the arithmetical operations on the real numbers are quite natural. To add two real numbers, we just add the corresponding terms of the Cauchy sequences defining the real numbers. For example, let $\{a_n\}$ be the sequence defined by the equation $a_n = 1 + 3/n$, which was shown earlier to be Cauchy, and let $\{b_n\}$ be the sequence used above to define the real number π. The sum of the real numbers $[\{a_n\}]$ and $[\{b_n\}]$ is the number determined by the sequence $\{a_n + b_n\}$ whose nth term is $a_n + b_n$:

(37) $\{a_n\}$: $4, 2.5, 2, 1.75, \ldots$

       $\{b_n\}$: $3, 3.1, 3.14, 3.141, \ldots$

       $\{a_n + b_n\}$: $7, 5.6, 5.14, 4.891, \ldots$

Here is the general definition:

DEFINITION 3.36.   If $\{a_n\}$ and $\{b_n\}$ are Cauchy sequences, then

(38) $[\{a_n\}] + [\{b_n\}] = [\{a_n + b_n\}]$.

There are two things that we must check in order to justify this definition. First, we must make sure that $\{a_n + b_n\}$ is a Cauchy sequence, since otherwise the notation $[\{a_n + b_n\}]$ makes no sense. And second, since we are working with equivalence classes, as usual we must show that the definition is unambiguous.

---

set are smaller than all numbers in the other. This alternative definition was first published by Dedekind, also in 1872, and the partitions of the rational numbers that are used in the definition are called *Dedekind cuts*. It can be shown that the two definitions of the real numbers are equivalent, in the sense that the real numbers as we have defined them are isomorphic to the real numbers defined in terms of Dedekind cuts. More details about Dedekind's definition, and a proof that the two definitions are equivalent, can be found in Strichartz (1995, 59–63).

THEOREM 3.37.
  (i) *If $\{a_n\}$ and $\{b_n\}$ are Cauchy sequences, then so is $\{a_n + b_n\}$.*
  (ii) *If $\{a_n\} \approx \{a'_n\}$ and $\{b_n\} \approx \{b'_n\}$, then $\{a_n + b_n\} \approx \{a'_n + b'_n\}$.*

Proof. See exercise 23. ∎

The definition of multiplication of real numbers is similar:

DEFINITION 3.38. If $\{a_n\}$ and $\{b_n\}$ are Cauchy sequences, then

$$(39) \quad [\{a_n\}] \cdot [\{b_n\}] = [\{a_n \cdot b_n\}].$$

The reader should be able to formulate a theorem similar to Theorem 3.37 that justifies this definition (see exercise 25). It is straightforward to use these definitions to prove the basic algebraic properties of the real numbers.

To define the ordering of the real numbers, suppose $\{a_n\}$ and $\{b_n\}$ are Cauchy sequences. We would like to say that $[\{a_n\}] < [\{b_n\}]$ if the location on the number line that $\{a_n\}$ approaches is to the left of the location that $\{b_n\}$ approaches. It is tempting to suggest that this will be true if and only if $a_n < b_n$ for all n, but this is wrong for two reasons. First of all, even if the location that $\{a_n\}$ approaches is to the left of the location that $\{b_n\}$ approaches, the numbers a_n and b_n may be close to these locations only when n is large; thus, for small n, it may not be true that $a_n < b_n$. Secondly, even if $a_n < b_n$ for all n, it is not necessarily true that the location that $\{a_n\}$ approaches is to the left of the location that $\{b_n\}$ approaches. They might be approaching the same location, with $\{a_n\}$ approaching that location from the left and $\{b_n\}$ approaching from the right. To rule this out, we will have to require that, for large n, a_n is *significantly* smaller than b_n. This leads us to formulate the following definition:

DEFINITION 3.39. Suppose $\{a_n\}$ and $\{b_n\}$ are Cauchy sequences of rational numbers. Then we define $[\{a_n\}] < [\{b_n\}]$ to mean that:

$$(40) \quad \exists k \in \mathbb{Z}^+ \exists N \in \mathbb{Z}^+ \forall n > N \left(a_n + \frac{1}{k} < b_n \right).$$

As usual, it is possible to prove that this definition is unambiguous (see exercise 26).

All of the familiar properties of the ordering of the real numbers can be proven from this definition. For example, we can prove the trichotomy law, which says that for any real numbers x and y, either $x < y$, $x = y$, or $y < x$ (see exercise 27). We can also prove that the real number line, unlike the rational number line, is a continuous line with no holes in it. To formulate this precisely, we need the concept of an upper bound for a set of real numbers.

DEFINITION 3.40.   Suppose $X \subseteq \mathbb{R}$. A real number b is called an *upper bound* for X if $\forall x \in X (x \leq b)$. If there is a smallest such number b, then it is called the *least upper bound* of X, and is denoted l.u.b.(X).

If a nonempty set X had an upper bound but no least upper bound, then there would be a hole in the number line at the boundary between the elements of X and the upper bounds of X. The following theorem guarantees that there are no such holes.

THEOREM 3.41 (Completeness of the real numbers).   *Every nonempty set of real numbers that has an upper bound has a least upper bound.*

Sketch of Proof. Suppose $X \subseteq \mathbb{R}$, $X \neq \varnothing$, and X has an upper bound. Let b be an upper bound for X, and let a be any element of X. Let a_1 be a rational number such that $a_1 < a$ and let b_1 be a rational number such that $b_1 > b$. (It is not hard to prove that such rational numbers exist.) Then a_1 is not an upper bound for X (since $a_1 < a$ and $a \in X$) and b_1 is an upper bound for X (since $b_1 > b$ and b is an upper bound). The least upper bound we are looking for will be somewhere between a_1 and b_1.

Let $c_1 = (a_1 + b_1)/2$, and note that c_1 is rational and that c_1 is halfway between a_1 and b_1. If c_1 is an upper bound for X, then let $b_2 = c_1$ and $a_2 = a_1$. If not, then let $a_2 = c_1$ and $b_2 = b_1$. Either way, b_2 is an upper bound for X and a_2 is not. The least upper bound of X will be between a_2 and b_2, and, since $b_2 - a_2 = (b_1 - a_1)/2$, this means we have narrowed down the search to an interval half as large as the one we started with.

Let $c_2 = (a_2 + b_2)/2$. If c_2 is an upper bound for X, then let $b_3 = c_2$ and $a_3 = a_2$. If not, then let $a_3 = c_2$ and $b_3 = b_2$. As before, b_3 is an upper bound for X and a_3 is not, and $b_3 - a_3 = (b_2 - a_2)/2$. Continue in this way to produce three infinite sequences of rational numbers, $\{a_n\}, \{b_n\}$, and $\{c_n\}$ (see Figure 3.1).

Although we will not go through the details, it is not hard to show now that all three sequences are Cauchy, and $\{a_n\} \approx \{b_n\} \approx \{c_n\}$. Let $x = [\{a_n\}] = [\{b_n\}] = [\{c_n\}]$. Then it can be shown that x is the least upper bound of X (see exercise 28). ∎

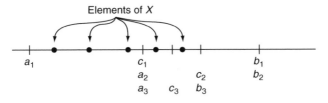

Figure 3.1

The completeness of the real numbers is used in the proofs of many of the theorems in calculus. As an example, we will consider a theorem about continuous functions from $\mathbb{R}$ to $\mathbb{R}$. Intuitively, a function $f : \mathbb{R} \to \mathbb{R}$ is said to be *continuous* at a real number x if $f(w)$ is close to $f(x)$ whenever w is a real number that is close to x. To turn this idea into a formal definition, we gradually make this statement more precise:

(41) If w is any number that is close to x, then $f(w)$ is close to $f(x)$.
$\forall k \in \mathbb{Z}^+$ (if w is any number that is sufficiently close to x, then $|f(x) - f(w)| < 1/k$).
$\forall k \in \mathbb{Z}^+ \exists j \in \mathbb{Z}^+$ (if w is any number such that $|x - w| < 1/j$, then $|f(x) - f(w)| < 1/k$).

Thus, we are led to the following definition of continuity:

DEFINITION 3.42. Suppose $f : \mathbb{R} \to \mathbb{R}$. Then, for any real number x, we say that f is *continuous at x* if

(42) $\forall k \in \mathbb{Z}^+ \exists j \in \mathbb{Z}^+ \forall w \in \mathbb{R}(|x - w| < 1/j \to |f(x) - f(w)| < 1/k)$.

We say that f is *continuous* if it is continuous at x for every $x \in \mathbb{R}$.

Another way to think about continuity of a function f is in terms of the graph of f. Just as the real numbers can be represented by a number line, the set $\mathbb{R} \times \mathbb{R}$ can be pictured as a plane, with the two entries in each ordered pair of real numbers specifying the horizontal and vertical positions of the corresponding point in the plane. Since a function f from $\mathbb{R}$ to $\mathbb{R}$ is a subset of $\mathbb{R} \times \mathbb{R}$, it corresponds to a certain subset of the plane, and this subset is called the *graph* of f. Since a pair $\langle x, y \rangle$ is an element of f if and only if $y = f(x)$, another way to say this is that the graph of a function f consists of all points $\langle x, y \rangle$ in the plane such that $y = f(x)$.

Intuitively, a continuous function is one whose graph is a continuous curve with no breaks or jumps in it.

THEOREM 3.43 (Intermediate Value Theorem).   *Suppose f is a continuous function from* $\mathbb{R}$ *to* $\mathbb{R}, u < v,$ *and* $f(u) < m < f(v)$. *Then there is a number w such that* $u < w < v$ *and* $f(w) = m$.

Figure 3.2 illustrates the Intermediate Value Theorem. In graphical terms, what the theorem says is that if a continuous curve is below the horizontal line $y = m$ at one point and above it at another, then it must cross the line somewhere in-between.

Sketch of Proof of Theorem 3.43.   Let $X = \{x \in \mathbb{R} : x < v$ and $f(x) < m\}$. Then X is nonempty, because $f(u) < m$ and so $u \in X$. Also, X has an upper bound, because every element of X is smaller than v, so v is an upper bound. Thus, by Theorem 3.41, X has a least upper bound. Let $w = \text{l.u.b.}(X)$. If $f(w) < m$, then, using the fact that f is continuous at w, we can find a number w' such that $w < w' < v$ and $f(w') < m$, so $w' \in X$, contradicting the fact that w is an upper bound for X. Similarly, if $f(w) > m$, then it can be shown that there is a number $w' < w$ such that w' is an upper bound for X, contradicting the fact that w is the *least* upper bound. Therefore, by trichotomy, we must have $f(w) = m$ (see exercise 29). ∎

As an application of Theorem 3.43, consider the function f defined by the equation $f(x) = x^2$. It can be shown that f is continuous, and clearly $f(1) = 1 < 2 < 4 = f(2)$. Therefore, by the Intermediate Value Theorem, there is some number w such that $1 < w < 2$ and

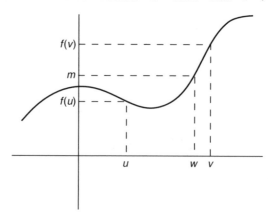

Figure 3.2

$f(w) = w^2 = 2$. Thus, although there is no rational number whose square is 2, there is a real number whose square is 2. In other words, by passing from the rational numbers to the real numbers, we have plugged the hole in the rational numbers at $\sqrt{2}$. (For another proof of the existence of $\sqrt{2}$, see exercise 30.)

As was mentioned in Chapter 2, Frege's idea of using one-to-one correspondences to compare the sizes of sets can be applied to infinite sets as well as finite sets. Applying this idea to the number systems we have defined leads to some surprising conclusions. For example, one might think that, since the set of rational numbers includes all of the natural numbers plus many more numbers, the sets of rational numbers and natural numbers must not be equinumerous. However, it turns out that they are equinumerous.

THEOREM 3.44. *The sets* $\mathbb{N}$ *and* $\mathbb{Q}$ *are equinumerous.*

Proof. The proof is based on the table in Figure 3.3, which extends infinitely far down and to the right and contains all of the rational numbers. The rational number p/q appears in the table in the column labeled p and the row labeled q.

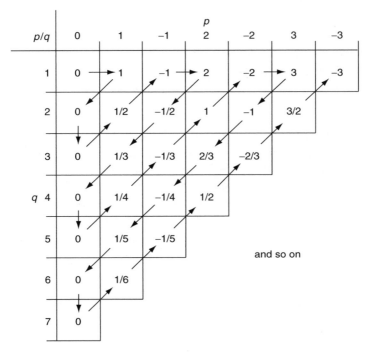

Figure 3.3

To create a one-to-one correspondence between $\mathbb{N}$ and $\mathbb{Q}$, we follow the arrows in this table, beginning at the 0 in the upper left, and list the numbers we come to. This gives us the following list of rational numbers:

$$0, 1, 0, 0, 1/2, \; -1, 2, \; -1/2, 1/3, 0, 0, 1/4, \; -1/3, 1, \; -2, 3,$$
$$-1, 2/3, \; -1/4, 1/5, 0, \ldots$$

Every rational number appears at least once in this list. Next, we eliminate repetitions by removing from the list any number that has already occurred earlier in the list. The result is the following list:

$$0, 1, 1/2, \; -1, 2, \; -1/2, 1/3, 1/4, \; -1/3, \; -2, 3, 2/3, \; -1/4, 1/5, \ldots$$

Finally, we pair the numbers in this list with the natural numbers, in order, to create the following set:

$$R = \{\langle 0, 0 \rangle, \langle 1, 1 \rangle, \langle 2, 1/2 \rangle, \langle 3, \; -1 \rangle, \langle 4, 2 \rangle, \langle 5, \; -1/2 \rangle, \langle 6, 1/3 \rangle,$$
$$\langle 7, 1/4 \rangle, \ldots\}.$$

It is not hard to see that R is a one-to-one correspondence between $\mathbb{N}$ and $\mathbb{Q}$, as required. The reader may be concerned that our definition of R has been somewhat informal, and relied on a diagram. However, it is possible to define R precisely without reference to a diagram, and it is possible to prove in ZFC that R is a one-to-one correspondence. Thus, $\mathbb{N}$ and $\mathbb{Q}$ are equinumerous. ∎

Sets that are equinumerous with $\mathbb{N}$ are called *denumerable*. A set is called *countable* if it is either finite or denumerable, and *uncountable* otherwise. Thus, Theorem 3.44 shows that $\mathbb{Q}$ is denumerable. It can also be shown that $\mathbb{Z}$ is denumerable. However, as our next theorem will show, $\mathbb{R}$ is not denumerable. Of course $\mathbb{R}$ is also not finite, so it is uncountable. This theorem was first proven by Cantor in 1874.

THEOREM 3.45.　*The sets $\mathbb{N}$ and $\mathbb{R}$ are not equinumerous.*

Proof. Suppose that $\mathbb{N}$ and $\mathbb{R}$ are equinumerous, and let R be a one-to-one correspondence between $\mathbb{N}$ and $\mathbb{R}$. Then R must be a set of ordered pairs of the form

$$R = \{\langle 0, r_0 \rangle, \langle 1, r_1 \rangle, \langle 2, r_2 \rangle, \langle 3, r_3 \rangle, \ldots\},$$

where $r_0, r_1, r_2, r_3, \ldots$ is a list of numbers in which every real number appears exactly once. We will now derive a contradiction by finding a real number that is missing from the list.

We begin by writing each of the numbers in the list in decimal notation. The decimal notation for r_i consists of an integer a_i, followed by a decimal point, and then followed by an infinite list of digits $d_0^i, d_1^i, d_2^i, \ldots$, each of which is between 0 and 9. In other words, in decimal notation we have $r_i = a_i.d_0^i d_1^i d_2^i \ldots$. We could imagine listing these decimal notations in a table:

Number	Decimal notation
r_0	$a_0.d_0^0 d_1^0 d_2^0 d_3^0 \ldots$
r_1	$a_1.d_0^1 d_1^1 d_2^1 d_3^1 \ldots$
r_2	$a_2.d_0^2 d_1^2 d_2^2 d_3^2 \ldots$
r_3	$a_3.d_0^3 d_1^3 d_2^3 d_3^3 \ldots$

and so on.

Now let e_0 be a digit that is different from d_0^0, let e_1 be a digit different from d_1^1, and so on. Since there are ten digits to choose from, it is easy to choose such digits. Let x be the number whose decimal expansion is $0.e_0 e_1 e_2 e_3 \ldots$. Since $e_0 \neq d_0^0$, the decimal expansions of x and r_0 disagree in the first position after the decimal point, so $x \neq r_0$. Similarly, the decimal expansions of x and r_1 disagree in the second position after the decimal point because $e_1 \neq d_1^1$, so $x \neq r_1$. In general, $x \neq r_i$ for any i, because $e_i \neq d_i^i$, and therefore the decimal expansions of x and r_i disagree at digit number $i + 1$. Thus x is a real number that is different from all of the numbers in the list $r_0, r_1, r_2, \ldots$, which contradicts the assumption that R is a one-to-one correspondence.[7] ∎

Notice that the digits of the number x in this proof were chosen to be different from the numbers $d_0^0, d_1^1, d_2^2, \ldots$, which appear in a diagonal line in the table above. For this reason the method used in the proof of Theorem 3.45 is called *diagonalization*. Diagonalization is a powerful method that can be used to prove many important theorems.

Since $\mathbb{N} \subseteq \mathbb{R}$, it is natural to conclude from Theorem 3.45 that $\mathbb{R}$ is larger than $\mathbb{N}$. In general, we will say that a set A is *larger than* a set B if B

---

[7] There is actually a small flaw in this proof, because it is possible for two different decimal expansions to represent the same real number; for example, $5.2999\ldots = 5.3000\ldots$. However, this happens only with decimal expansions that end in infinitely many 0's or 9's, so the proof can be fixed by making sure that $1 \leq e_i \leq 8$.

is equinumerous with a subset of A, but B is not equinumerous with A. In terms of this ordering on the sizes of infinite sets, it turns out that the denumerable sets are the smallest infinite sets:

THEOREM 3.46.   *Every infinite set has a denumerable subset.*

Proof. Suppose A is an infinite set. Then A is certainly not the empty set, so we can choose an element $a_0 \in A$. Since A is infinite, $A \neq \{a_0\}$, so we can choose some $a_1 \in A$ such that $a_1 \neq a_0$. Similarly, $A \neq \{a_0, a_1\}$, so we can choose $a_2 \in A$ such that $a_2 \neq a_0$ and $a_2 \neq a_1$. Continuing in this way, we can recursively choose $a_n \in A$ such that $a_n \notin \{a_0, a_1, \ldots, a_{n-1}\}$. Now let $R = \{\langle 0, a_0 \rangle, \langle 1, a_1 \rangle, \langle 2, a_2 \rangle, \ldots\}$. Then R is a one-to-one correspondence between $\mathbb{N}$ and the set $\{a_0, a_1, a_2, \ldots\}$, which is a subset of A. Therefore, A has a denumerable subset.   ∎

Although this proof seems convincing, it cannot be formalized using the set theory axioms that we have listed so far. The axioms we have discussed guarantee the existence of sets that are explicitly specified in various ways – for example, as the set of all subsets of some set (Axiom of Power Sets), or as the set of all elements of some set that have a particular property (Axiom of Comprehension). But the proof of Theorem 3.46 does not specify the one-to-one correspondence R completely, because it does not specify how the choices of the elements $a_0, a_1, a_2, \ldots$ are to be made. To justify the steps in this proof, we need an axiom guaranteeing the existence of sets that result from such arbitrary choices:

AXIOM OF CHOICE.   Suppose F is a set of sets such that $\varnothing \notin F$. Then there is a function C with domain F such that, for every $X \in F$, $C(X) \in X$.

The function C is called a *choice function* because it can be thought of as *choosing* one element $C(X)$ from each $X \in F$. To see how the Axiom of Choice (often abbreviated AC) can be used to justify the steps in the proof of Theorem 3.46, suppose A is an infinite set, and let $F = \{X \in \mathcal{P}(A) : X \neq \varnothing\}$. By AC, we can let C be a choice function for F. Now we can use C to specify how the elements $a_0, a_1, a_2, \ldots$ in the proof of Theorem 3.46 are to be chosen, as follows. Let $X_0 = A$. Then $X_0 \in \mathcal{P}(A)$ and $X_0 \neq \varnothing$, so $X_0 \in F$. Thus, we can let $a_0 = C(X_0) \in X_0$. Next, let $X_1 = \{x \in A : x \neq a_0\}$, and note that $X_1 \in \mathcal{P}(A)$ and $X_1 \neq \varnothing$, so $X_1 \in F$. Let $a_1 = C(X_1)$. In general, for any n, we let $X_n = \{x \in A : x \neq a_0 \land x \neq a_1 \land \ldots \land x \neq a_{n-1}\}$, and we let $a_n = C(X_n)$. The set X_n is the set of all elements of A that would be acceptable values for a_n, and the choice function C decides for us which of these values we should choose.

The Axiom of Choice was first advanced by Zermelo in 1904, and he included it in his 1908 axiomatization of set theory.[8] The axiom generated a great deal of controversy when Zermelo introduced it, although many of those who objected to it had themselves used, in their mathematical work, reasoning similar to that used in the proof of Theorem 3.46 above. In 1938, the Austrian logician Kurt Gödel (1906–78) proved that, if the axioms of ZF (Zermelo–Fraenkel set theory *without* AC) do not lead to any contradictions, then no contradictions are introduced by the inclusion of AC as an additional axiom. In 1963, the American mathematician Paul Cohen (1934–) proved that, if the axioms of ZF do not lead to any contradictions, then AC cannot be proven in ZF. Today, most mathematicians accept AC, and much of modern mathematics depends on AC for its correctness. For instance, another example of a theorem whose proof requires AC is the trichotomy law for infinite cardinalities: for any two infinite sets A and B, either A and B are equinumerous or one of them is larger than the other.

Having established that the denumerable sets are the smallest infinite sets, and that the set of real numbers is not denumerable, it is natural to wonder if $\mathbb{R}$ is the second smallest infinite set. The *continuum hypothesis* (CH) is the assertion that there is no infinite set that is smaller than $\mathbb{R}$ but not denumerable. It was conjectured by Cantor in 1878 but, despite considerable effort, he was unable to prove it. Today, as a result of the work of Gödel in 1938 and Cohen in 1963, we know that, if the axioms of ZF do not lead to any contradictions, then CH is neither provable nor disprovable in ZFC. We will have much more to say about such "undecidable" questions in Chapter 7.

Exercises

1. Prove Theorem 3.9. (Hint for part (iv): You may find it helpful to prove part (v) first, so that you can use mathematical induction. Then use mathematical induction to prove the following two properties of natural numbers:

---

[8] Several other mathematicians had earlier recognized the possibility of using arbitrary choices in defining sets, but Zermelo appears to have been the first to advocate the acceptance of such reasoning. For a thorough discussion of this and other issues surrounding AC, see Moore (1982).

    (a) For every natural number x, every element of x is also a subset of x.

    (b) For every natural number x, $x \notin x$.

    Finally, use (a) and (b) to prove part (iv) of Theorem 3.9.)

2. Justify Definition 3.10 by proving that there is a unique function $P : \mathbb{N} \times \mathbb{N} \to \mathbb{N}$ such that, for every $m \in \mathbb{N}$,

    (1) $P(\langle m, 0 \rangle) = m$, and

    (2) for every $n \in \mathbb{N}$, $P(\langle m, S(n) \rangle) = S(P(\langle m, n \rangle))$.

    (Hint: For every natural number k, let us say that a function $f : \mathbb{N} \times S(k) \to \mathbb{N}$ is a *partial plus up to k* if for every $m \in \mathbb{N}$,

    (1') $f(\langle m, 0 \rangle) = m$, and

    (2') for every $n \in k$, $f(\langle m, S(n) \rangle) = S(f(\langle m, n \rangle))$.

    First show that for every $k \in \mathbb{N}$ there is a unique partial plus up to k. Then let F be the set of all partial plusses, let $P = \cup F$, and prove that P has the required properties.)

3. Prove the cancellation property for addition of natural numbers: for all natural numbers m, n, and k, if $m + k = n + k$ then $m = n$. (Hint: Let m and n be arbitrary natural numbers, and then prove $\forall k (m + k = n + k \to m = n)$ by induction on k.)

4. Prove that for all natural numbers m and n, $S(m) + n = S(m + n)$. (Do *not* use the commutative law for addition. This fact will be used to prove the commutative law.)

5. Prove that addition of natural numbers is commutative: for all natural numbers m and n, $m + n = n + m$. (Hint: Use exercise 4.)

6. Prove that multiplication of natural numbers is associative: for all natural numbers m, n, and k, $(m \cdot n) \cdot k = m \cdot (n \cdot k)$.

7. Prove that multiplication of natural numbers is commutative: for all natural numbers m and n, $m \cdot n = n \cdot m$. (Hint: Imitate exercise 5.)

8. Prove the trichotomy law for natural numbers: for all natural numbers m and n, either $m < n$, or $m = n$, or $n < m$. (Hint: Let m be an arbitrary natural number, and then prove by induction that $\forall n (m < n \lor m = n \lor n < m)$. You may find it useful to prove first that, for every natural number k, if $k \neq 0$ then there is some natural number j such that $k = S(j)$.)

9. Suppose R is an equivalence relation on A. As in the text, for each $x \in A$, let $[x]_R = \{y \in A : xRy\}$, and let $A/R = \{[x]_R : x \in A\} = \{X \in \mathcal{P}(A) :$ for some $x \in A$, $X = [x]_R\}$. Prove the following facts about A/R:

    (a) For every $X \in A/R$, $X \neq \varnothing$.

    (b) For every $X, Y \in A/R$, if $X \neq Y$ then $X \cap Y = \varnothing$.

    (c) $\cup(A/R) = A$.

10. Complete the proof of Theorem 3.18 by showing that $\sim$ is reflexive and symmetric.

11. Prove Theorem 3.24. (Hint: You may find it useful to prove as an intermediate step that $\langle ac + bd, ad + bc \rangle \sim \langle a'c + b'd, a'd + b'c \rangle$.)

12. State and prove the theorem needed to show that Definition 3.25 is unambiguous.

13. Prove that, for all integers x and y, $(x + y) - y = x$.

14. Prove that, for every integer x, $x + (-1) \cdot x = 0$. (First identify what integers the numerals "-1" and "0" stand for.)

15. Prove Theorem 3.28.

16. State and prove the theorems necessary to justify Definition 3.30.

17. Prove the distributive law for rational numbers: for all rational numbers x, y, and z, $x \cdot (y + z) = (x \cdot y) + (x \cdot z)$.

18. Suppose we tried to define a new operation $\oplus$ on the rational numbers as follows: for all integers a, b, c, and d, $[\langle a, b \rangle] \oplus [\langle c, d \rangle] = [\langle a + c, b + d \rangle]$. Explain why this "definition" is unacceptable. Your explanation should include an example illustrating the ambiguity of the "definition."

19. Let $\{a_n\}$ be defined as follows:

$$a_n = \begin{cases} 0, & \text{if } n \leq 1000, \\ 1, & \text{if } n > 1000. \end{cases}$$

Let $\{b_n\}$ be the sequence in which $b_n = 1$ for every n. Show that $\{a_n\}$ is a Cauchy sequence and that $\{a_n\} \approx \{b_n\}$.

20. Suppose $\{a_n\}$ is a Cauchy sequence, and let $\{b_n\}$ be the sequence resulting from leaving out the first 100 terms of $\{a_n\}$; in other words, $b_1 = a_{101}, b_2 = a_{102}$, and in general, $b_n = a_{100+n}$. Show that $\{b_n\}$ is a Cauchy sequence and $\{a_n\} \approx \{b_n\}$.

21. Suppose that, for each $n \in \mathbb{Z}^+$, $d_n \in \{0, 1, 2, \ldots, 9\}$. For each $n \in \mathbb{Z}^+$, let

$$s_n = \frac{d_1}{10} + \frac{d_2}{100} + \cdots + \frac{d_n}{10^n}.$$

In other words, s_n is the number whose representation in decimal notation is $0.d_1 d_2 \ldots d_n$. Prove that $\{s_n\}$ is a Cauchy sequence. (Of course, $[\{s_n\}]$ is the real number whose representation in decimal notation is $0.d_1 d_2 \ldots$.)

22. Complete the proof of Theorem 3.34 by showing that $\approx$ is reflexive and symmetric.

23. Prove Theorem 3.37.

24. Prove that if $\{a_n\}$ is Cauchy then it is bounded; in other words, there is some positive integer K such that, for all $n \in \mathbb{Z}^+$, $|a_n| < K$.

25. State and prove the theorem needed to justify Definition 3.38. (Hint: To prove that $\{a_n \cdot b_n\}$ is Cauchy, you will need to work with the formula $|a_m \cdot b_m - a_n \cdot b_n|$. Begin by rewriting it as follows: $|a_m \cdot b_m - a_n \cdot b_n| = |a_m \cdot b_m - a_m \cdot b_n + a_m \cdot b_n - a_n \cdot b_n|$. Now use the triangle inequality and exercise 24.)

26. State and prove the theorem needed to justify Definition 3.39.

27. Prove the trichotomy law for real numbers: for any real numbers x and y, either $x < y$, $x = y$, or $y < x$. (Hint: Suppose $x = [\{a_n\}]$ and $y = [\{b_n\}]$. Assume that $x \neq y$, which means that $\{a_n\} \not\approx \{b_n\}$, and prove that either $x < y$ or $y < x$.)

28. Complete the proof of Theorem 3.41 by showing that $\{a_n\}$, $\{b_n\}$, and $\{c_n\}$ are Cauchy, $\{a_n\} \approx \{b_n\} \approx \{c_n\}$, and $x = [\{a_n\}] = [\{b_n\}] = [\{c_n\}]$ is the least upper bound of X.

29. Complete the proof of Theorem 3.43 by showing that if $f(w) < m$ then there is some real number w' such that $w < w' < v$ and $f(w') < m$, and if $f(w) > m$ then there is some real number $w' < w$ such that w' is an upper bound for X.

30. For every positive integer n, let a_n be the unique rational number i/n with the property that

$$\left(\frac{i}{n}\right)^2 < 2 < \left(\frac{i+1}{n}\right)^2.$$

For example, $a_5 = \frac{7}{5}$, because $\left(\frac{7}{5}\right)^2 < 2 < \left(\frac{8}{5}\right)^2$. Show that $\{a_n\}$ is a Cauchy sequence and $[\{a_n\}]^2 = 2$.

4

Intuitionism

By the end of the last chapter, the reader might conclude that Frege's project, although it got off to something of a rough start, has after all been shown to be realizable. Working within what one could consider a naive theory of extensions, Frege's logical system is indeed vulnerable to Russell's Paradox; but once a more sophisticated approach to sets (like that of ZFC) is put in its place, paradox can be kept at bay and the reduction of mathematics to logic can proceed unthreatened by contradiction.

Yet hardly anyone thinks that the logicist project, as Frege envisioned it, has been carried through. Why not? Is the answer that the reduction of arithmetic, say, to set theory, conceived of as a theory of extensions, fails to count as a logicist reduction because this theory does not qualify as logic? Extensions are not logical entities but, rather, mathematical ones – this answer might continue – and so Frege's project was ill-conceived from the start.

But why *not* consider extensions logical entities? True, no such entities are typically treated of in a course on logic. By that token, however, Frege's entire project would be doomed from the start: for natural numbers are not usually taken to be logical either, even though the success of Frege's logicism depends on showing that they are. To deny logical status to extensions simply because they are not commonly viewed as part of logic is to place too much importance on pre-theoretical intuitions about what "seems logical" and what does not.

But this does raise the question of why Frege thought the theory of extensions was logical. The answer must be that this theory has the requisite generality, for Frege took generality to be the hallmark of logic, which comprises "those laws of thought that transcend all particulars."[1] Extensions are completely general in part because every concept has one.

---

[1] *Begriffsschrift*, p. 5.

In fact, the problem with reducing arithmetic to Frege's theory of extensions is not that these entities fail somehow to be logical in nature; it is rather, as Russell's Paradox shows, that the theory of these extensions is inconsistent. By contrast, the problem with reducing arithmetic to ZFC is not that this theory is inconsistent (as far as we know, it is not), but rather that it is not completely general, and for this reason not logical.

ZFC fails to exhibit the requisite generality in a number of different ways. In the first place, the theory makes particular existential assertions about sets, the entities with which it concerns itself. Most obviously, there is the Axiom of Infinity, which asserts the existence of a set that contains infinitely many sets. This is a very specific assertion about the universe of sets, one that detracts from the generality of the theory. The axioms of ZFC do not all exhibit the kind of presuppositionlessness with regard to the universe of sets that would be needed to warrant being considered logical.

But secondly, and more basically, ZFC fails to be general purely by dint of being a theory of sets, for sets – unlike extensions – fail to correspond to all concepts. Consider, for instance, the concept *set* itself. We saw in the last chapter, in Theorem 3.3, that we can prove in ZFC that there is no set corresponding to this concept: there is no set of all sets. There is, of course, an extension corresponding to the concept, namely the class that contains all sets. But as we noted, Russell's Paradox is in effect transformed in ZFC into a proof that this class is a proper class, that is, not a set. The truly general notion is that of class: corresponding to every concept there is a class (some classes will be sets, the others proper classes). But ZFC is not a theory about classes. It is a theory about sets, about entities which – whatever one calls them – are not completely general, in that there are concepts to which no such entity corresponds.

ZFC fails, then, to have the generality that would warrant deeming it logical. Consequently, a reduction of mathematics to it (such as was sketched in the previous chapter) cannot constitute a vindication of Frege's logicist program. Whatever mathematical interest there might be in such a reduction, it cannot resolve the philosophical questions scouted in earlier chapters. It is the generality of logic that promises that a reduction to it would have some philosophical point. A reduction of mathematics to ZFC, by contrast, deposits all our philosophical problems on the doorstep of a theory that is no better (arguably, far worse) equipped to answer them than is, say, arithmetic.

In sum, we know that Frege's theory of extensions, though logical, is inconsistent, and that ZFC, though a consistent theory of sets (so far as we know) to which mathematics can be reduced, lacks the requisite generality to be considered appropriately logical. Whence our frustration? What is it about Frege's goal that makes it so elusive? Is there some reason why we cannot arrive at a consistent theory of extensions to which mathematics can be reduced, which is also thoroughly general in some philosophically relevant sense?

In order to approach this question from a helpful angle, we must be more explicit about the perspective thus far adopted with respect to mathematical reality. We can best appreciate it by considering first what our ordinary attitude is to statements about the natural world. Consider, for example, the claim

(1)   Every star has at least one planet orbiting it.

What is our stance with respect to the truth of (1)? Most of us will say that we do not know whether (1) is true or false. Perhaps no one knows yet whether (1) is true or false. But what is important in this connection is that most of us will insist, assuming we have clearly specified what constitutes a star, a planet, and so on, that (1) is determinately true or false. It will be true just in case each and every star in the universe has at least one planet orbiting it, and it will be false just in case there is at least one star that lacks an orbiting planet. Reality is one way or the other. We do not know which at present, but we do know that matters are either as (1) describes or they are not, and that (1) is correspondingly true or false.

This compelling view of the natural world is called *realism*. It involves a picture of Nature as independent of us and determinate: it is a quite definite way, regardless of our ability to know which way it is. Nature makes every meaningful statement about it true or false, although we might never know which. Consequently, a hallmark of our realist stance toward the natural world is that we are prepared to assert the *Law of the Excluded Middle* for all statements about it:

(2)   For all statements *S*, either *S* is true or not-*S* is true.

Our insistence that either (1) is true or its negation is follows from our acceptance of (2) for statements about the natural world.

Most mathematicians adopt such a realist posture toward the mathematical world. For them, the collection of natural numbers is in the relevant respect no different from the collection of stars in the universe. Recall Hardy's credo, quoted in Chapter 1: "I believe that mathematical reality lies outside us, that our function is to discover or *observe* it, and

that the theorems which we prove, and which we describe grandilo-
quently as our 'creations,' are simply our notes of our observations."[2]

Consider, for example, the following assertion:

(3)   Every even number greater than 2 is the sum of two primes.

This is known as Goldbach's Conjecture, and is one of the most famous
unsolved problems in number theory. It has been computer-tested for
billions of numbers and no counter-example has been found, but so far
no proof of it has been discovered either. Nevertheless, the mathematician on the street will insist that either (3) is true, or else there is a
particular even number greater than 2 that is not the sum of two primes.
This mathematician will go on to insist that not only do we not know
now which is the case, but we might never know. Still, regardless of our
lack of knowledge, the mathematical world, the world of natural
numbers in particular, is either one way or the other. The common
mathematician is disposed to apply the Law of the Excluded Middle to
mathematical assertions. And indeed, in the case of Goldbach's Conjecture for instance, one might well wonder how it could not apply. For
consider the extension of *even number greater than 2*: it is the collection
of all and only those entities that fall under this concept. This collection is
a determinate, infinite totality: $\{4, 6, 8, \ldots\}$. Surely, regardless of what
anyone might believe or come to know, either every element of that
totality is the sum of two primes, or at least one is not; in the first case,
the claim is true, in the second its negation is. To doubt this line of
thought might even seem unintelligible to many. Certainly, most mathematicians, Frege included, have not doubted it.

Nevertheless, realism in mathematics leads to certain pressures that
some have found too uncomfortable to bear. One such we have touched
on: it concerns what the realist says about the concept *set*. Consider the
extension of this concept; let us call it *U*, the class of all sets. As we have
already seen, *U* is not a set according to ZFC: if it were, we could, using
the Axiom of Comprehension, form the set of all those elements of *U* that
are not members of themselves. But such a set does not exist, on pain of
contradiction. Therefore, *U* is a proper class. Now while this might at
first seem a rather unexceptionable state of affairs, it can easily come to
appear quite mysterious. For recall the motivation that was sketched in
the last chapter for the ZFC hierarchy of sets. One begins with Ø at the
first stage, and then, at each successive stage, one adds the power set of
the collection of sets formed at, or before, the previous stage. At any

[2]   Hardy (1967, 123–4).

given stage, the entities obtained are sets. In the beginning, this process results in only a finite number of sets: after each of the stages V_0, V_1, V_2, and so on, we have succeeded in collecting together finitely many sets. The first stage that collects together infinitely many sets is only reached "after" all these: V_ω is what one gets when one collects together all the sets constructed at all the infinitely many prior stages. It is natural to feel uncomfortable at this point. But if we object that one cannot complete an infinite process, that there is no "after" when dealing with the infinite, the realist will respond that while one cannot in practice do this, in principle one could. Imagine now that someone asks the realist whether one can collect together all sets formed at all stages, to form the set of all sets. The classical set-theorist will of course respond that this cannot be done. But if we press, and ask why this particular infinite gathering-together of previous stages to create a set is impossible whereas others are not, the set-theorist can offer nothing except the reply that in the one case taking unions across stages results in contradiction, while in the other it does not. To say that in the one case, but not in the other, there are "too many" stages to collect across is of no help: in both cases, an infinite number of stages are involved, and there is nothing more to say about what makes the one infinite process contain "too many" steps beyond the fact that contradiction rears its head if we imagine completing a process of that length. There is no explanation here at all.

This situation can seem thoroughly unsatisfactory. A classical mathematician might nevertheless believe that, uncomfortable as it might appear, it is simply forced upon one by the above proof that U, the class of all sets, is not itself a set. What step in the proof, after all, can one so much as question? A critic of realism will respond that it is precisely the tacit assumption of realism, an assumption so basic and pervasive that the classical mathematician might not even recognize it as such, that is at the root of the problem. The proof that there is no set of all sets proceeds by *reductio ad absurdum* from the assumption that

(4)   There is a set U that contains all sets,

in effect employing Russell's Paradox to generate a contradiction. The classical mathematician takes it for granted that there is a collection U of all sets, and concludes from the *reductio* that (4) is to be rejected *because U is not a set*. The realist assumes that mathematical entities, such as sets, are part of an objective and determinate mathematical reality. Each concept divides that reality into two: those entities that fall under the concept and those that do not; in particular, corresponding to the concept *set*, there is a fixed collection that contains each and every set. According to the critic of realism, this assumption is what forces the classical mathematician to reject (4) by making mysterious brute distinctions between collections that are sets and those that are not. Once this assumption is isolated, we can begin to make out an alternative. For another approach would be to assume instead that every collection is a set, and then to conclude from the reductio that (4) must be rejected *because there is no collection that contains all sets*. So construed, the argument begins by assuming that there is a collection U that contains all sets, and then shows that in fact U does not by producing a set that is not in U, namely the set R that, by the Axiom of Comprehension, contains all those sets in U that are not members of themselves.

We are not, after all, forced to view the argument as the realist does. But if we choose not to, what precisely are we saying about concepts like *set*? Did we not introduce U as the extension of the concept *set*? If it turns out that U cannot contain all sets, then were we not wrong to identify it as the extension of *set*? The critic of realism will insist that what this proof teaches us is that there is a distinction between those concepts whose extensions are fixed, and those concepts that are *indefinitely extensible*.[3] An indefinitely extensible concept is one whose extension cannot be completely determined: any collection of entities that falls under the relevant concept can be extended. Given any collection of entities that falls under an indefinitely extensible concept, we can describe an entity which falls under that concept but which is not contained in that collection. Some concepts have extensions that can be fixed, the concept *prime less than 100* for instance. But this is not so for all concepts. In particular, the concept *set* is indefinitely extensible: given any collection of sets, we can define a set that does not belong to the original collection. In fact, the critic of realism will contend, this is the more natural way of viewing the application of Russell's Paradox: given any collection of sets M, we can define another set which is nevertheless not in M – namely, the set of all sets in M that are not members of themselves.

The view that some concepts are indefinitely extensible goes hand-in-hand with a conception of mathematical reality that is at odds with realism. For it is plausible that if the mathematical universe is determinate, then to any well-defined concept there must correspond a fixed collection of each and every entity in that universe which falls under that concept. The indefinite extensibility of some concepts signals, then,

---

[3]   The term is due to the English Philosopher Michael Dummett (1925–); see, for example, Dummett (1991, 317). He notes there that the idea goes back at least to Russell (1905, 135–64; see esp. 144).

that mathematical reality is not fully determinate. Not every intelligible question that can be asked about it has an answer. Truths, on this view, are no less objective or necessary than they are according to realism: once we see that they obtain, there can be no question but that they do, indeed must, hold. But it is a confusion then to infer that there is a reality that already now settles all possible questions. The following might provide a helpful analogy. Consider some random process – say, for the sake of the argument, the tossing of a coin. That the coin just tossed came up heads is an objective truth, but that does not mean that there is already now a fact about what each subsequent toss of the coin will yield. Likewise, it is a fact that M, any given collection of sets, can be extended by adding to it the set of all sets in M that are not elements of themselves, for this set does not belong to M; and this extended set M' can in turn be extended by adding the set of all sets in M' that are not members of themselves; and so on. But it is a confusion to think that in the mathematical universe there is a set that contains the elements of M together with every single set that would be generated by this process were it to continue forever. This anti-realist conception of mathematics is sometimes called *constructivism*.[4] In what follows, we shall consider a particular constructivistic approach known as *intuitionism*.[5]

It seems, then, as if we have a choice. Either we take there to be a collection that contains all sets, in which case we must distinguish between totalities that are sets and those that are not; or we treat all collections as sets, in which case we must distinguish between those concepts that are indefinitely extensible and those that are not. The first option has its home within realism, whose conception of a determinate mathematical reality seems to underwrite the belief that there is a totality that contains all sets. By contrast, the second option is more naturally associated with a view of mathematical reality as not fully determinate. The clash between these two pictures of the mathematical universe is confirmed by the fact that this choice cannot be evaded. This is the lesson of Russell's Paradox: we cannot simultaneously assume that

---

[4]   Constructivistic objections to realism in mathematics were raised in the nineteenth century, if not before, thus predating Russell's Paradox. This paradox, however, makes the cost of a full-fledged mathematical realism especially clear.

[5]   Naturally, the word "intuitionism" has been used to name a number of different, though closely related, theories. The term itself was coined by the Dutch mathematician L. E. J. Brouwer (1881–1966), who pioneered the approach in the early twentieth century. According to him, our intuition, in a Kantian sense of the term, of time forms the basis for all of mathematics. Our development of intuitionism follows rather different lines, some of which Brouwer would have opposed.

there is a collection that contains all sets and also treat that collection as if it were a set.

Putting the matter this way reveals why Russell's Paradox goes to the heart of Frege's project, and not merely to some particular attempt to realize it. For Frege both espouses a realist attitude toward mathematics and seeks to reduce it to logic. This combination of views leads Frege to treat extensions as belonging to an independently existing and determinate universe, and also as being fully general. But generality is just what a theory lacks if it treats of only *some* extensions, which is precisely the restriction that Russell's Paradox forces upon modern set theory in the context of its realistic interpretation of the set-theoretic universe.

Realism thus leads the set theorist to distinguish between kinds of extensions: some are sets, others not. Intuitionism compels one to distinguish between kinds of concepts: some are indefinitely extensible, others not. Which distinction to favor, then?

The intuitionist is in no doubt. We have seen that the distinction drawn by the ZFC set-theorist between sets and proper classes appears unmotivated: Why should some collections of sets be sets themselves and others not? The distinction may appear to be vindicated by its offering a solution to Russell's Paradox. But in fact, ZFC does not provide a solution to the Paradox, which concerns collections construed most generally, so much as choose to talk about something else, namely sets. The contradiction is not explained by the distinction between sets and proper classes. On the contrary, this distinction is given its content by the unexplained threat of contradiction: we have no paradox-independent criterion for drawing the line between sets and proper classes. To insist on this distinction in order to avoid paradox is, as Dummett has put it, "to wield the big stick, not to offer an explanation."[6] For just this reason, there is something fundamentally mysterious about the subject matter of ZFC, for its contours are shaped by a contradiction for which no explanation is given.

By contrast, the intuitionist continues, appeal to the distinction between indefinitely extensible concepts and other concepts has some explanatory force in this context, for that distinction can be motivated independently of the Paradox. How? Intuitionists hold that it is unavoidable once one insists, as they do, that infinite collections cannot be *actual*. For intuitionists, who believe that mathematical reality is not fully fixed and independent of us but instead remains to be determined through our mathematical activity, the only kinds of collections that can intelligibly be spoken of as actual are ones whose construction it would make sense

[6]   Dummett (1994, 22).

to imagine ourselves completing, if only in principle. Any given finite process is one we can in principle complete, and so any finite collection can intelligibly be treated as actual. By contrast, we cannot, even in principle, complete an infinite process: for a process to be infinite just is for it never to come to an end, for it to make no sense imagining its completion. Consequently, a totality cannot sensibly be viewed as actually infinite. Its infinitude is always *potential*: it consists really of a principle which, given any collection (and of course, the only collections one can actually be given are finite ones), will generate an element of the totality not contained in that collection. Consider, for example, the natural numbers. According to the intuitionist, the only actual collections of natural numbers are finite ones: {0}, {0, 1}, {0, 1, 2}, and so on. The totality of all natural numbers does not exist as an actual, completed collection of numbers. Rather, its infinitude consists in the fact that given any collection of natural numbers, we can generate a natural number that is not in that collection, say by adding 1 to the largest element of that collection. For one who holds that the infinite is always only potential, never actual, there corresponds to each infinite collection an indefinitely extensible concept: for the infinitude of the collection can now only consist in the fact that, given any collection of entities that falls under the corresponding concept, we can characterize another entity which falls under that concept but which is not contained in that collection. Once the completed infinite is rejected, that some concepts are indefinitely extensible and others are not is no more mysterious than that some concepts have infinite extensions and others do not.

The realist, however, rejects this framework and embraces the completed infinite. What humans can in principle do has no bearing on the nature of mathematical reality. Whether we can complete an infinite process is one thing, the realist will insist, and whether there are infinite collections of mathematical entities is quite another. As a consequence, the notion of an indefinitely extensible concept strikes the realist as misguided. A concept carves up the world of mathematical entities into those that fall under it and those that do not. And since the denizens of that world exist and have the properties they do independently of us, there simply must be a collection that contains all and only those entities that fall under the concept. Even if any collection that we could in principle generate would fail to contain all entities that fall under the concept, there still exists a collection that contains just those entities. From the realist's perspective, the whole idea of an indefinitely extensible concept is erected upon a confusion between the capacities of human beings on the one hand, and mathematical facts on the other. Our capacities determine how much knowledge we can acquire about the mathematical world, but it is a mistake to think that they at all determine how matters stand in that world.[7]

Before proceeding to some illustrations, it is worth pausing to make a couple of observations about what is *not* at issue between these two different perspectives. In the first place, their fundamental disagreement is not over the existence of mathematical entities. Both the classical mathematician and the intuitionist agree, for instance, that the number 2 exists. They even agree that infinite totalities exist, although of course they differ in their conceptions of infinity. The basic difference between these two positions concerns not whether there is a mathematical reality but, rather, the nature of the reality that is said to exist, in particular whether it is completed and determinate. Secondly, the issue that divides realists and intuitionists does not concern the constitution of mathematical entities themselves. While realism is often associated with the view that the objects of mathematics are abstract, and intuitionism with the view that they are in some fashion mental constructions, these associations miss what fundamentally separates the two positions. For there is no intimate connection between the ontological constitution of mathematical entities – the "stuff," so to speak, of which they are made – and the nature of the reality of which they are a part, in particular its completedness and determinacy. For instance, one might hold that natural numbers are in some sense mental entities, but yet insist that the collection of them is a completed infinite one.[8] And, conversely, someone might view the natural numbers as abstract entities, but at the same time take the concept *natural number* to be indefinitely extensible.[9]

Let us now turn to some examples which may help to highlight the real differences between these two frameworks. We shall first consider one of the most famous proofs in mathematics, Euclid's proof that there are infinitely many prime numbers. It proceeds as follows: Assume that there are only finitely many primes, p_1, p_2, ..., p_n. Let $q = (p_1 \cdot p_2 \cdots p_n) + 1$. Now q is not divisible by p_1, for the division leaves a remainder of 1; and by the same token, q is not divisible by p_2, or by..., or by p_n. But since every number is a unique product of primes

[7]   There are also intermediate positions that neither eschew all actual infinite totalities, as intuitionism does, nor accept them all, as realism does. For an articulation and defense of the view that only denumerable collections form completed infinite totalities, see Velleman (1993).

[8]   Dedekind perhaps held this; see his "The nature and meaning of numbers'" Dedekind (1888); Ewald (1996).

[9]   This position is advanced by Michael Dummett; see, for instance, Dummett (1978).

(this has to be separately established), there must be some prime other than $p_1, p_2, \ldots, p_n$ that divides q. QED.

The classical mathematician sees this as a proof that the extension of the concept *prime* is a completed infinite totality. That this extension is a completed totality is part and parcel with realism. And that this completed totality is of infinite size follows from the fact that it is larger than any finite collection of entities. Intuitionists, by contrast, see here no vindication of the notion of the completed infinite. What the proof actually proves, according to them is, rather, that the concept *prime* is indefinitely extensible: given any collection of primes, we can find a prime number that is not contained in it. In fact, this proof gives us instructions for doing so: first, find the corresponding number q, then list its divisors, and finally check each one for primality. This is a finite process, and the proof guarantees that it will terminate in a prime other than the ones with which we began. The infinitude of prime numbers is potential, and just consists in the fact that any given collection of them can be extended to a larger one. The classical mathematician talks as if this process of arriving at ever larger collections of primes could somehow be terminated, with the completed infinite collection of primes standing at its terminus; but the intuitionist insists that this is not established by the proof, and indeed that this picture makes no sense at all.

We saw, in the previous chapter, Cantor's proof that the collection of real numbers is nondenumerable. Again, while both the classical mathematician and the intuitionist accept the proof, they offer substantially different accounts of what the proof actually proves. The realist comes to the proof with the assumption that the concept *real number* determines a completed extension: a collection that contains all the real numbers and nothing else. Before Cantor's proof, one might have thought that this collection, being infinite, is of the same size as the infinite collection of natural numbers. What Cantor showed is that this is incorrect: the completed totality of natural numbers cannot be mapped 1–1 onto the completed totality of real numbers. Although infinite, the totality of reals is of a different size from that of the natural numbers; in short, there are different sizes of infinity.

In sharp contrast, the intuitionist insists that this interpretation depends on much mathematical ideology, for nowhere in the proof is there mention of different sizes of completed infinite collections. What the proof actually shows is that, given any function from the natural numbers to the reals, we can find a real number that is not a value of that function. The proof provides us with a method which, given a function for enumerating real numbers, finds a real not so enumerated. Just as infinitude for the intuitionist means only that there is a principle for extending any given collection to a more inclusive one, so nondenumerability means that any given principle for enumerating collections can itself be extended. According to the intuitionist, the classical mathematician takes these facts about extendability and incoherently reifies them by talking about different kinds of completed collections: finite, denumerable, nondenumerable.

The reader at this point might well be wondering whether there is any substance to the debate between the classical mathematician and the intuitionist. We know that the one talks of a fixed mathematical reality, of concepts determining completed totalities, and of the actual infinite; and the other of the indeterminacy of this reality, of the indefinite extensibility of some concepts, and of the potential infinite. But do these differences result in any real disagreement? Might "realist-speak" be a mere *façon de parler*, a vivid, object-oriented way of describing facts about finite processes? Might both positions accept the same proofs, after all, but perhaps merely talk about them in different ways?

A proof consists in a finite sequence of inferences. A disagreement about whether to accept a proof arises should there be a conflict regarding the validity of one of its steps of reasoning. Most generally, an inference will be valid just in case the correctness of asserting its premises guarantees the correctness of asserting its conclusion. A conflict regarding the validity of an inference must then be traceable to a disagreement about the conditions in which some statement can be correctly asserted. To determine, therefore, whether the classical mathematician and the intuitionist will substantively disagree, we must examine the conditions in which they take mathematical statements to be correctly assertable, that is, we must consider what each might mean when making a mathematical assertion. Let us begin with a statement about particular numbers: "17 and 19 are twin primes," where two numbers are *twin primes* just in case they are both prime and their difference is equal to two. Whether or not one takes a realist perspective on the mathematical world, there will be no disagreement about when this statement can be asserted. For given any two numbers, a *finite* computation will reveal whether they are twin primes. Consequently, the fundamental difference between the realist and the anti-realist concerning the infinite could not lead to any disagreement here. More generally, the classical mathematician and the intuitionist will agree about any statement whose assertability hinges on a finite calculation.

Matters stand differently, however, when we move to statements that involve quantification over an infinite domain. Let us consider, for example, the following assertion and its partial logical analysis:

(5)   There are infinitely many twin primes.

(6) $\forall x \exists y (y > x \wedge y$ is prime $\wedge y + 2$ is prime$)$.

How will a realist interpret this claim? The classical mathematician understands the universal quantifier "$\forall x$" to range over all elements of a completed infinite totality, in this case the completed collection of natural numbers: $\{0, 1, 2, 3, \ldots\}$. What the statement says is that every element of that completed infinite totality has the property that there are twin primes greater than it; only if this actual infinitary circumstance obtains will assertion of (6) be correct. The intuitionist, on the other hand, finds that this cannot possibly be what (6) means, since the whole notion of a completed infinite totality makes no sense. The collection of natural numbers does not exist as a completed totality, and there is likewise no corresponding actual infinitary circumstance in which each of these numbers is smaller than some twin primes. Rather, the concept *natural number* is an indefinitely extensible one: given any collection of natural numbers, we can generate another natural number not contained in it. From this perspective, what (6) states is that no matter which natural number we are given, we can produce twin primes greater than it. The realist interprets the statement as making a claim about the elements of some completed infinite totality. The anti-realist, by contrast, understands it to be about the entities that could in principle be generated by some operation.[10]

With these contrasting meanings in mind, let us consider whether circumstances might arise in which the classical mathematician would judge it legitimate to assert (6) but the intuitionist would not. Imagine that both the realist and the intuitionist believe that it is not the case that it is not the case that (6) holds; in other words, that they both assert the statement's double negation:

(7) $\neg\neg\forall x \exists y (y > x \wedge y$ is prime $\wedge y + 2$ is prime$)$.

What would they each infer from this? According to the realist, the collection of natural numbers is a determinate, independently existing, infinite totality. One of the following two possible conditions must therefore obtain: (i) every element in that collection is smaller than

---

[10]   The difference in meaning here will be missed if one thinks of the collection of these generatable entities as a completed infinite totality. For the intuitionist, no given collection can encompass an operation's infinite output; put adverbially, its total output does not exist actually, but only potentially.

some twin primes; or (ii) it is not the case that every element in that collection is smaller than some twin primes. If (ii) fails to hold, then (i) must hold. Statement (7) says that (ii) fails to obtain. The classical mathematician will therefore assert (6) on the basis of (7).

The intuitionist, however, will not. An anti-realist does not believe that an infinite collection of natural numbers actually exists, and will dismiss any reasoning based on this assumption as confused. What (7) tells us is that we cannot assert the negation of (6). But that does not guarantee, the intuitionist will insist, that we *can* assert (6). We could assert (6) if we knew that we can assert either (6) or its negation. But this remains to be shown; the illusion that it need not be shown, that it is obviously correct, is fed by realist assumptions about the determinacy and completedness of the mathematical universe. Put otherwise, intuitionists grant that acceptance of (7) entails that we cannot assert the negation of (6), but that we are barred from doing so, they insist, does not by itself furnish us with a way of showing that (6) holds, of proving that given any natural number we could produce some twin primes greater than it. If one does not take the natural numbers to be a completed infinite totality, then in order to accept a universal statement of the form "Every natural number has property *P*" one would need a guarantee that each natural number that can be generated can be seen to have property *P*. We might naturally view such a guarantee as a foolproof method for establishing that any natural number that can be produced can be shown to have *P*. And just knowing that (7) is the case does not furnish us with such a method.

What this shows is that the inference

$$(8) \quad \frac{\neg\neg X}{X} \quad \text{(double negation elimination)}$$

is not in general accepted by the intuitionist.[11] This is of course incomprehensible from a realist perspective: if one views mathematical statements as describing an independently existing and determinate reality, then if "$\neg X$" fails to describe that reality correctly, it must be that "X" does correctly describe it. After all, the belief that mathematical reality is completed and determinate is part and parcel with the belief that, for any

---

[11]   The notation

$$\frac{P_1 P_2 \ldots P_n}{Q}$$

means that if it is correct to assert P_1 and correct to assert P_2 and . . . and correct to assert P_n, then it is also correct to assert Q.

statement, either it describes that reality correctly or its negation does. The anti-realist rejects this perspective, and so distinguishes between knowing that ¬X fails to be a correct assertion and knowing that X is a correct assertion. Here, then, we see how the realist's and the intuitionist's interpretations do actually lead to disagreements about what is a legitimate inference.[12]

We can already see that intuitionists will not always accept the way classical mathematicians argue: mathematical argument proceeds by inference, and here we have a fundamental disagreement over the validity of a particular form of reasoning, double negation elimination. We need now to determine just where these disagreements will arise. We are immediately confronted by the fact that there are infinitely many possible inference forms. How can we determine those that both realists and intuitionists will agree are valid, and those about whose validity they will disagree? A valid argument is one that transmits correct assertability; that is, one with the property that if its premises are correctly assertable then its conclusion must be as well. Thus, what we require is a general account of the conditions of correct assertability for the classical mathematician and a similar account for the intuitionist. With these in hand, we can examine particular inference forms to see whether realists and intuitionists agree about their validity.

For the classical mathematician, what makes the assertion of a statement correct is that the mathematical world, taken to be independently complete and determinate, is as the statement claims it to be. We know when it would be correct to assert a statement – in an important sense, we know what it means – when we know which conditions must obtain in the independently existing, determinate mathematical world in order for the statement to be true. This is often put by saying that the classical conditions for correct assertability are *truth conditions*. The familiar recursive definition of truth for classical first-order logic can thus be viewed as a specification of how each logical connective contributes to the conditions of correct assertability of statements in which it appears. For instance, this definition specifies that a universal quantification $\forall x Fx$ will be true under interpretation $\mathfrak{J}$ if and only if every object in $\mathfrak{J}$'s

---

[12]   It should be noted that the intuitionist will not reject every inference of the form (8). In particular, if A is not a statement that makes an infinitary claim, say our earlier "17 and 19 are twin primes," then the corresponding application of double negation elimination will be deemed valid. We shall return to this below, after we have said more about the intuitionist's understanding of negation. The point here is that, according to the intuitionist, double negation elimination is not legitimate for every A, and hence is not in general a valid principle of inference.

universe of discourse is contained in the extension that $\mathfrak{J}$ assigns to the predicate *F*. According to the classical mathematician, this fixes the conditions of correct assertability of any universal quantification. Our conclusions above about the classical interpretation of (6) were just a particular application of this. It is by reference to assertability conditions, so specified, that the classical mathematician judges the validity of an inference form.

We must now do something comparable for the logical connectives as understood by the intuitionist. For each logical connective, we shall offer a specification of the conditions in which a statement governed by that connective would be correctly asserted. We shall thereby specify the intuitionistic conditions of correct assertability for a statement of any given logical form. If a statement is correctly assertable, then there must be some actual state of affairs by virtue of which it is correctly assertable. We have seen that these circumstances cannot be classical truth conditions, for the intuitionist finds those that involve the completed infinite fundamentally unintelligible. Because the only actual circumstances recognized by the intuitionist are finite, the correct assertability conditions for statements must be finitary; they must be such that it makes sense to imagine a human being – possibly idealized, but not beyond the bounds of intelligibility – recognizing them. If we bear in mind that a mathematical proof must be finite, then one natural way of implementing the intuitionist's perspective is just to identify the assertability conditions of statements with proofs themselves. Because this analysis of a statement's assertability conditions proceeds by reference to what would count as a proof of it, we might call it an analysis in terms of *proof conditions*.

If we treat a statement's assertability conditions as its meaning, then we might say that from the realist's perspective, although meaning and proof-condition are distinct, the meaning of a statement dictates what counts as a proof of it. From this perspective, therefore, the intuitionist's proposal appears peculiar in that it removes from the picture just what one would naturally look to as determining what a statement's proof could be. For the intuitionist, however, the proposal's excision of meanings that we can make no sense of is precisely its cardinal virtue. The intuitionist fails to grasp the meanings the realist associates with mathematical statements, their realist truth conditions, and *a fortiori* fails to understand how those meanings serve to determine what should count as proofs. The intuitionist's proof-conditional approach, by contrast, eliminates any substantial gap between the meaning of a statement and the conditions for its justification: the meaning associated with a mathematical statement by the intuitionist just is its proof condition. And the worry about intelligibility is now also allayed: because of proof's finite nature, the clarity of proof conditions on that score cannot be gainsaid.

We must now be more specific about what these intuitionistic proof conditions are. For the time being, we shall restrict our discussion to arithmetical statements, that is, to statements about the natural numbers. A logically atomic arithmetical statement is simply an equation or an inequality, and a proof of such a statement consists of a finite computation that confirms it. Thus, for example, the primitive computation that the sum of 3 and 2 is 5 is, for the intuitionist, a proof of the equation "$3 + 2 = 5$". We now recursively specify what constitutes a proof of a logically complex arithmetical statement:

(9) (i)   A proof of $X \land Y$ is a pair of proofs, the one a proof of X and the other a proof of Y.
     (ii) A proof of $X \lor Y$ is a proof either of X or of Y.
     (iii) A proof of $X \to Y$ is an operation that transforms every proof of X into a proof of Y.
     (iv) A proof of $\neg X$ is a proof of $X \to 0 = 1$.
     (v)   A proof of $\forall x Fx$ is an operation that, for every numeral n, produces a proof of Fn.[13]
     (vi) A proof of $\exists x Fx$ is a proof, for some specific numeral n, of Fn.

A statement is correctly assertable, according to the intuitionist, if there is a proof of it.

For example, let us consider how (9) applies to a particular statement, say, (6) repeated here:

(6) $\forall x \exists y (y > x \land y$ is prime $\land y + 2$ is prime$)$.

First, apply (9)(v), which says that a proof of (6) is an operation that, given a numeral n, produces a proof of "$\exists y(y > n \land y$ is prime $\land y + 2$ is prime)." Now apply (9)(vi), which tells us that a proof of (6) is an

---

[13] By "numeral" we mean a term constructed in the usual manner using "0" and the symbol for the successor function: either "0", or a finite sequence of "S"s followed by "0". Numerals provide us with a way of referring to any natural number. Furthermore, every numeral refers to a natural number. In the following chapter, we shall consider quantification over domains other than the natural numbers. With respect to some such domains, the ways we have of describing their elements will allow descriptions that fail to refer to any element in the domain. As we shall then see, this necessitates a slight elaboration in our analysis of the meaning of quantification.

operation that, given a numeral n, produces a numeral m and a proof of "$m > n \wedge m$ is prime $\wedge m + 2$ is prime." Eventually, we apply the base case, so that a proof of (6) turns out to be an operation that, given a numeral n, produces a numeral m such that if one carries out the computations involved in checking that $m > n$, that m is prime, and so on, they all come out the right way. This is just the result one would expect on the basis of our earlier informal discussion of the intuitionistic meaning of (5). A number of observations are now in order.

First, note that the difference between the classical mathematician and the intuitionist is not that the one makes infinitary claims while the other refrains from doing so. They both find infinitary assertions intelligible; they differ, however, in the assertability conditions that they associate with such claims.

Secondly, it would be a mistake to think intuitionism differs from realism in judging statements that have not yet been proved to be unintelligible. One need not have proved a statement in order to understand it; for the intuitionist, it suffices to know what would count as a proof of it. For example, (6), although thus far undecided, is intelligible to the intuitionist because (9) specifies precisely what its proof conditions are.

The contrast between a truth-conditional analysis of assertability conditions and a proof-conditional one is nevertheless substantial, and we are now in a position to characterize one broad difference. Because classical assertability conditions will in general involve reference to how things are in some completed infinite totality and because humans are not capable of directly checking that such an infinitary condition obtains, it follows that a statement could be classically assertable even though it might not be possible for us to determine that it is.[14] The situation is otherwise for the intuitionist. Intuitionistic assertability conditions are finite proofs. Hence, if such a condition obtains, it is always possible for us to be presented with it and to recognize it as a proof of the appropriate kind. Consequently, it is always possible for us to ascertain that a statement is intuitionistically assertable, if it is.

It is natural to question this. One might well acknowledge that intuitionistic assertability conditions are finite, and hence that it is possible for them to be given to us in a way in which we cannot be given classical assertability conditions, and yet still question whether it is always possible for us to recognize that we have indeed been given a proof: to take in a proof is one thing, to determine that it is a proof of a given statement

---

[14] This inference assumes that it is not always the case that whenever a statement's truth conditions hold, we can prove that they do. Not all classical mathematicians believe this, however. We will return to this in Chapter 6.

is another. Consider, for example, the intuitionist's account of the proof of a universal statement. According to (9)(v), such a proof is an operation, which is of course a finite entity. But in order to determine whether a presented operation is indeed a proof of some given universal statement, we have to determine whether the operation transforms every given numeral into a proof of the corresponding instance. Could it not be the case that a presented operation effects this transformation but that we cannot tell that it does? Likewise for the proof of a conditional, as explained in (9)(iii): Might it not be impossible for us to discover that an operation does in fact transform every proof of the antecedent into a proof of the consequent?

It is quite correct that, while intuitionistic assertability conditions are indeed finite entities, the properties they have in virtue of which they are proofs are infinitary. And it is also true that if we take the collection of numerals, or of proofs, to be a completed infinite totality, then it might not be possible for us to know that a given operation is a proof, if it is. But of course, the intuitionist's intention behind (9) will be thoroughly missed if one conceives of the collection of numerals, or of proofs, in this way. To accept such a conception of these collections would just be to reintroduce a version of the realist's actual infinite that the intuitionist rejected as unintelligible. Rather, the concepts *numeral* and *proof* must be viewed as indefinitely extensible: no actual collection encompasses every entity that falls under either concept.[15] Once we view the matter thus, there is no basis for thinking that an operation might have the relevant infinitary property although this fact be forever unknowable by us.

This response might in turn lead to a different worry about the intuitionist's conception. If the concepts *numeral* and *proof* are indefinitely extensible, then quantification over numerals and proofs cannot be understood as the realist understands it. That is, if one is to be faithful to the intentions of the intuitionist, the very quantifiers in the explanatory clauses of (9) must be understood intuitionistically, not classically. But then it seems as if (9) could be properly understood only by someone who did not need to understand it: if (9) can be interpreted as intended only if one construes its quantifiers intuitionistically, one must wonder how it can serve to convey a concept of intuitionistic quantification to someone who does not already have it.

Whatever we are to make of this, it should be noted that the situation is no different for the classical mathematician: the familar semantics for classical logic is given in a meta-language whose quantifiers must them-

---

[15]   In Chapter 6, in our discussion of Gödel's Incompleteness Theorems, we shall return to the idea that the concept *proof* is indefinitely extensible.

selves be interpreted classically. For instance, when we say that $\forall x F x$ is true under an interpretation just in case every object in that interpretation's universe of discourse is contained in the extension the interpretation assigns to F, "every" is intended to be understood as ranging over a completed infinite totality of entities. In neither the classical nor the intuitionistic case could such verbal accounts convey, through being understood as intended, the target understanding to someone who lacked it: for they could only be so understood by someone who already possessed the relevant conception. Both these construals of the logical vocabulary of our language are sufficiently fundamental as to rule out any noncircular way in which they can be explicitly described. In fact, it can appear rather mysterious how we acquired these conceptions in the first place (if acquisition is even the correct way to view the matter). We shall not pursue these matters further here, and we simply reiterate that (9) needs to be interpreted in the right way in order for it to express the intended intuitionistic interpretation of the logical connectives.

The best way of understanding the import of (9) is to return to the very project for which we introduced it: an examination of inference forms with a view to deciding whether intuitionists and realists would agree about their validity. The classical mathematician accepts the Law of the Excluded Middle, as we have seen. We can view this as acceptance of the following inference form:

(10) $$\frac{}{Y \lor \neg Y}.$$

What this means is that from *nothing* one can infer $Y \lor \neg Y$; the conclusion, that is, can be inferred from anything. The classical mathematician takes this to be valid, of course: every circumstance is one in which $Y \lor \neg Y$ is true. But the validity of (10) means something different for the intuitionist: it would require that we can always generate a proof of $Y \lor \neg Y$. A proof of this is, by (9)(ii), a proof of Y or a proof of $\neg Y$. But there is no reason to think that, given any statement, we can prove it or refute it. If one thought that for every mathematical statement there already exists either a proof of it or a refutation of it, although it might be impossible for us, even in principle, to ascertain which, then this would indeed justify the inference – but at the cost of a realist conception of the totality of proofs, which the intuitionist eschews. For the intuitionist, if there is a proof of a statement or a refutation of it, then it must be possible for us to determine in principle which it is; that is, we should in principle be able to produce the proof or produce the refutation. But for most statements, we are not in such a position: it is simply not the case, for example, that we can in principle prove Goldbach's Conjecture or refute it. Hence, the intuitionist will not affirm the validity of (10).

We can put this last observation by saying that the intuitionist does not accept the Law of the Excluded Middle. However, it would be a mistake to infer from this that there is some "middle" state, as it were, that the intuitionist thinks a mathematical statement can be in: that of being neither provable nor refutable. For an intuitionist, to prove that a statement is not provable is to show that any proof of it can be transformed into an absurdity, which, according to (9)(iv), is just to refute it. A statement may well be neither proved nor refuted; this is, in fact, the present situation with regard to Goldbach's Conjecture or with regard to (6). But it makes no intuitionistic sense to imagine that a statement could be shown to be forever resistant both to proof and to refutation: for the intuitionist, one refutes a sentence by showing that it can never be proved.[16]

Consider now the following inference:

$$(11a) \quad \frac{X}{\neg\neg X}.$$

This is valid classically. To see if it is also valid intuitionistically, it will help to rewrite its conclusion like this: $(X \rightarrow 0 = 1) \rightarrow 0 = 1$. A proof of this is an operation that takes any proof of $X \rightarrow 0 = 1$ and transforms it into a proof of $0 = 1$. Can we construct such an operation on the assumption that we have a proof of X, the premise of (11a)? We can. To see this, let us also assume that we have a proof of $X \rightarrow 0 = 1$, and show that we can now generate a proof of $0 = 1$. Recall that to have a proof of $X \rightarrow 0 = 1$ is just to have an operation that takes a proof of X, which we have assumed given, and transforms it into a proof of $0 = 1$. This argument is therefore valid for the intuitionist.

Its converse, however, is not:

$$(11b) \quad \frac{\neg\neg X}{X}.$$

---

[16] We are assuming here that to refute a sentence, i.e., to prove that it implies "$0 = 1$", is to show that it cannot be proved. This identification relies on the assumption that it is impossible to prove "$0 = 1$". The intuitionist will insist on this: the only thing that would count as a proof of "$0 = 1$" is a direct comparison of the two numbers 0 and 1 that results in a finding that they are equal. And, in fact, a direct comparison shows that they are not equal.

We have already seen that the intuitionist will not accept this inference. And now we can see why more clearly. In general, there is no reason to think that given an operation that transforms any refutation of X into a proof of $0 = 1$, which we might view as a proof that X will never be refuted, we can arrive at a proof of X. The intuitionist, unlike the classical mathematician, does not take a statement and its double negation to be equivalent: they are not always guaranteed to be assertable in the same circumstances.

We noted earlier that matters would be different if one were guaranteed that either X is provable or refutable. This knowledge, together with the proof that X cannot be refuted, can in effect be transformed into a proof of X itself. In other words,

$$(12) \quad \frac{X \vee \neg X \quad \neg\neg X}{X}$$

is valid intuitionistically. To have a proof of $X \vee \neg X$ is either to have a proof of X or to have a proof of $\neg X$. The proof of $\neg\neg X$ is a proof that $\neg X$ cannot be proved. So, the proof of $X \vee \neg X$ must, in fact, be a proof of X.

One consequence of the general failure of double negation elimination is that *reductio ad absurdum* must be used with care. While the intuitionist accepts both

$$(13) \quad \frac{X \rightarrow (Y \wedge \neg Y)}{\neg X},$$

the form of argument employed in the proof of Theorem 1.1, and

$$(14) \quad \frac{\neg X \rightarrow (Y \wedge \neg Y)}{\neg\neg X},$$

the following is not intuitionistically valid:

$$(15) \quad \frac{\neg X \rightarrow (Y \wedge \neg Y)}{X}.$$

Even if we are given that any refutation of X can be transformed into a proof of a contradiction, that does not by itself tell us how to prove X. In general, the intuitionist will only accept arguments by *reductio* to a *negated* conclusion, unlike the classical mathematician, who makes no such distinction.

Other familiar methods of argument will also have to be reassessed. For example, argument by cases:

$$(16) \quad \frac{X \rightarrow Y \quad \neg X \rightarrow Y}{Y}$$

is not intuitionistically valid, and hence our proof of Theorem 1.2 would not be accepted by an intuitionist. Given proofs of the two premises, we might not yet be able to construct a proof of Y: that might have to await either a proof of X or its refutation. In other words, the following would be an intuitionistically acceptable version of argument by cases:

$$(17) \quad \frac{X \rightarrow Y \quad \neg X \rightarrow Y \quad X \vee \neg X}{Y}.$$

The classical mathematician can detect no difference between (16) and (17), since the Law of the Excluded Middle is a classical logical truth. For the intuitionist, on the other hand, it is a substantive claim, and any appeal to it requires a justification.

According to the intuitionistic interpretation of the connectives, familiar truth-functional equivalences will also have to be treated more carefully. For instance, the classical mathematician affirms the equivalence between $X \rightarrow Y$ and $\neg X \vee Y$. The intuitionist accepts the implication in only one direction, however. Thus,

$$(18a) \quad \frac{\neg X \vee Y}{X \rightarrow Y}$$

is intuitionistically valid. If we have a proof of the premise, then we have either a proof of $\neg X$ or a proof of Y. Assume now that we are given a proof of X. It follows that our first proof could not have been a proof of $\neg X$: for that would be a proof that X cannot be proved, and we are assuming that we have just been given a proof of X. So our first proof must in fact have been a proof of Y. The inference form is therefore a valid one.

By contrast,

$$(18b) \quad \frac{X \rightarrow Y}{\neg X \vee Y}$$

is not intuitionistically acceptable: given an operation that transforms any proof of X into a proof of Y, we might not be able to extract from it either a proof of ¬X or a proof of Y. Again, if we could in principle either generate a proof of X or generate its refutation, then the inference would be valid. In the first case, the operation given in the proof of the premise can be applied to the proof of X to yield a proof of Y, and so of the conclusion; and in the second, we are given a proof of ¬X straight-away, and so we are again done. But of course the intuitionist does not agree that $X \lor \lnot X$ is correctly assertable for all statements.

It should be clear by now that we cannot take for granted that the classically valid equivalences will also be intuitionistically acceptable. Let us consider a few more examples involving the conditional. In classical logic, a conditional is logically equivalent to its contrapositive. This is only half true in intuitionistic logic, according to which

$$(19a) \quad \frac{X \to Y}{\lnot Y \to \lnot X}$$

is valid, but

$$(19b) \quad \frac{\lnot Y \to \lnot X}{X \to Y}$$

is not. To see that (19a) is intuitionistically valid, it might help to express it as follows:

$$(19a') \quad \frac{X \to Y}{(Y \to 0 = 1) \to (X \to 0 = 1)}.$$

Let us assume that we have a proof of its premise (call it "o_1"). Do we now have a proof of the conclusion? Assume that we are given also a proof of the antecedent of the conclusion (call it "o_2"). Can we now produce a proof of $X \to 0 = 1$; that is, an operation that will transform any proof of X into a proof of $0 = 1$? We can, for assume that we are given a proof of X; using o_1, we can construct a proof of Y, to which o_2 can be applied to yield a proof of $0 = 1$, as desired. Inference (19b) is not valid, however: in general, an operation that transforms any refutation of Y into a refutation of X does not yield an operation that transforms any proof of X into a proof of Y.

The intuitionist draws distinctions where the classical mathematician sees none. Some of these are rather subtle. For instance, the intuitionist takes the following to be valid:

$$(20) \quad \frac{X \to \lnot Y}{Y \to \lnot X}.$$

This can be seen either directly, or by noting that the inference can be indirectly justified as follows. Assume that we are given a proof of Y, the conclusion's antecedent; in the light of (11a), we can produce a proof of $\neg\neg Y$. But from a proof of (20)'s premise and (19a), we can produce a proof of $\neg\neg Y \rightarrow \neg X$. And therefore we can construct a proof of $\neg X$. By contrast,

(21) $\quad \dfrac{\neg X \rightarrow Y}{\neg Y \rightarrow X}$

is not valid. Readers might try to convince themselves of this directly, but it is interesting to see why an indirect justification analogous to that just employed for (20) does not succeed. So, assume that we are given a proof of the conclusion's antecedent; that is, of $\neg Y$. From a proof of (21)'s premise and (19a), we can produce a proof of $\neg Y \rightarrow \neg\neg X$. Now if (11b) were valid intuitionistically, then we could construct a proof of X. But since it is not, the indirect proof fails.

Let us turn now to consider four basic inferences involving quantifiers and negation. An examination of their status in intuitionistic logic will help to convey the flavor of the intuitionist's interpretation of the quantifiers. We begin with

(22a) $\quad \dfrac{\exists x \neg Fx}{\neg \forall x Fx}$.

Can we transform any proof of its premise into a proof of its conclusion? Assume we are given a proof of its premise; by (9)(vi), we have a proof of a particular substitution instance, say $\neg Fk$. The question is whether there is an operation that transforms every proof of $\forall x Fx$ into a proof of $0 = 1$. There is: given a proof of the universal quantification, apply it to generate a proof of Fk; this, together with the proof of $\neg Fk$, yields a proof of $0 = 1$. The inference form is consequently intuitionistically valid.

Again, the classical mathematician accepts the converse as well:

(22b) $\quad \dfrac{\neg \forall x Fx}{\exists x \neg Fx}$.

But on the intuitionist's construal of the connectives, this is not valid. We might be able to show that every proof of the universal quantification leads to an absurdity without being able to describe a particular entity that can be proved to lack the relevant property, which is what a proof of the argument's conclusion requires.

The other familiar pair of inferences are these:

(23a) $\dfrac{\forall x \neg Fx}{\neg \exists x Fx}$

and

(23b) $\dfrac{\neg \exists x Fx}{\forall x \neg Fx}$.

Of course, the classical logician takes both to be valid. It may be surprising to learn that the intuitionist does as well. We shall leave the intuitionistic justification of (23a) to the reader, and consider only that for (23b). Assume, then, that we have a proof of its premise, an operation that carries a proof of any substitution instance of $\exists x Fx$ into a proof of $0 = 1$. That is, we have an operation that, for any numeral k, carries a proof of Fk into a proof of $0 = 1$. But this would be an operation that transforms every numeral k into a proof of $Fk \rightarrow 0 = 1$, and such an operation is just a proof of $\forall x \neg Fx$, the argument's conclusion. This asymmetry of judgments concerning the pairs (22) and (23) underscores the need to think carefully about what each argument is claiming when its logical vocabulary is interpreted intuitionistically.

Let us conclude by surveying some basic inferences involving quantifiers and the conditional. Consider the following four pairs of arguments:

(24a) $\dfrac{P \rightarrow \forall x Fx}{\forall x(P \rightarrow Fx)}$,   (24b) $\dfrac{\forall x(P \rightarrow Fx)}{P \rightarrow \forall x Fx}$;

(25a) $\dfrac{P \rightarrow \exists x Fx}{\exists x(P \rightarrow Fx)}$,   (25b) $\dfrac{\exists x(P \rightarrow Fx)}{P \rightarrow \exists x Fx}$;

(26a) $\dfrac{\forall x Fx \rightarrow P}{\exists x(Fx \rightarrow P)}$,   (26b) $\dfrac{\exists x(Fx \rightarrow P)}{\forall x Fx \rightarrow P}$;

(27a) $\dfrac{\exists x Fx \rightarrow P}{\forall x(Fx \rightarrow P)}$,   (27b) $\dfrac{\forall x(Fx \rightarrow P)}{\exists x Fx \rightarrow P}$.

For the classical logician, these are all valid inferences. This is not so, however, for the intuitionist: (25a) and (26a) fail to be valid intuitionistically (see exercise 1). At the root of these failures is the very strong interpretation that the intuitionist gives to existential quantifications: in order for an existential statement to be correctly assertable, one must be able in principle to describe the entity that has the property in question. Of course, putting the matter this way is misleading. This is a "strong" interpretation only from the perspective of the classical mathematician. From the intuitionist's point of view, this is just what an existential assertion means. The intuitionist views the classical interpretation of the existential quantifier not as weaker, but instead as unintelligible.[17]

It is plain, then, that many of the logical inferences the classical mathematician relies on will not be acceptable to the intuitionist. In fact, we have already seen that the intuitionist will reject at least one of the proofs, that of Theorem 1.2, with which we began in Chapter 1, since it employs the intuitionistically invalid inference (16). This might have led the reader to wonder about the status of the proof of Theorem 1.3, which relies on an inference that is not on its face logical in nature, namely mathematical induction. We know that the logicist justifies induction by logically deducing it from a definition of the concept *natural number*. This would not be acceptable to the intuitionist, however, for the logicist definition is impredicative: as the logicist understands it, it contains a second-order quantifier that ranges over the very concept being defined. Such an understanding is of a piece with a view of the concept *natural number* as an element in some determinate universe of concepts. Viewed as picking out the already existing concept *natural number*, the impredicative definition is, as the American philosopher W. V. Quine (1908–2000) has remarked, "not visibly more vicious than singling out an individual as the most typical Yale man on the basis of averages of Yale scores including his own."[18] But the intuitionist rejects this conception of the universe of concepts, and so rejects such an impredicative characterization.

This raises the question of how the intuitionist defines the natural numbers. From the perspective of intuitionism, a completely adequate characterization says that the natural numbers are just those objects obtainable from 0 by repeatedly applying the successor operation, S. We might regard this characterization as giving two rules for generating natural numbers:

(i) 0 is a natural number, and
(ii) if n is a natural number, then so is $S(n)$.

---

[17] The intuitionistic invalidity of some of these inferences might lead the reader to wonder whether in intuitionistic logic every statement is equivalent to one in prenex normal form (i.e., one in which no quantifier is in the scope of any logical expression except a quantifier), as is the case in classical logic. The answer is no: $\neg\forall x Fx$, for example, is not intuitionistically equivalent to any prenexed formula.
[18] Quine (1969, 243).

The natural numbers are those, and only those, objects that are generated by these rules. The restriction that *only* objects generated by rules (i) and (ii) are numbers is often referred to as the *extremal clause* of the definition.

This analysis will appear quite inadequate to classical mathematicians. In particular, they will object that there is nothing in it that prevents a nonfinite iteration of the second generating rule, (ii); and such iteration must be ruled out, the objection continues, for otherwise there is no guarantee that the definition will capture only the natural numbers, and nothing else. The intuitionist's definition will thus appear at best elliptical to the classical mathematician: for it must be understood that (ii), the second rule of the definition, permits only *finite* iteration of the successor operation to yield natural numbers. Furthermore, it will not do to secure this restriction on the application of (ii) by stating explicitly that only *finite* iterations of it are permitted, where "finite iteration" is interpreted to mean "iteration n times, for some natural number n." For this would render the analysis blatantly circular. The classical mathematician, for this reason, might well offer instead the following as a friendly amendment to the intuitionist's analysis: delete the inexplicit extremal clause, and secure its intent by specifying that induction is valid. For the classical mathematician, the requirement that induction is a valid means of forming generalizations about the elements of some collection guarantees that nonnumbers will be excluded from the collection. For the predicate "natural number" applies to 0, and applies to $S(n)$ if it applies to n, and therefore it must apply to every element of a collection for which induction is valid. Of course, this entailment presupposes that "natural number" (or a predicate of a piece with it) is taken to be well-defined: the validity of induction does not articulate the intention of the extremal clause unless induction is understood impredicatively, as a generalization over a pre-existing domain of predicates that includes the very predicate being defined. As we have observed, such impredicativity is unacceptable from the intuitionistic point of view; consequently, the intuitionist will not consider induction to be an adequate replacement for the extremal clause.

In fact, the intuitionist will find the friendly amendment not only unhelpful, but unnecessary as well. From the intuitionistic viewpoint, no intelligible but unwanted possibility has yet been described that would require changes to, or replacement of, the extremal clause: since the intuitionist accepts the potential infinite, and not the completed infinite, the idea of applying the generation rules an infinite number of times is simply unintelligible. The feared possibility is one that only makes sense given a conception of the infinite that the intuitionist categorically rejects; it is pointless to rule out explicitly what is in fact an impossible application of the relevant rules.

A question remains for intuitionism, however: Given its characterization of the natural numbers, how is induction to be justified? Clearly, the intuitionist must offer some account of induction, if its use is to be permitted in mathematical proofs. If we express induction as the following inference:

$$(28) \quad \frac{F0 \quad \forall x(Fx \;\rightarrow\; FS(x))}{\forall x Fx},$$

then the task becomes that of establishing the intuitionistic validity of (28). This is not difficult to do. Assume that we are given a proof of $F0$, a proof of $\forall x(Fx \rightarrow FS(x))$, and a numeral k. Here is how to construct a proof of Fk. First, applying our proof of $\forall x(Fx \rightarrow FS(x))$ to the numeral 0, we produce a proof of $F0 \rightarrow FS(0)$. Given our proof of its antecedent, we can then produce a proof of $FS(0)$. Next, applying again our proof of $\forall x(Fx \rightarrow FS(x))$, this time to the numeral $S(0)$, we construct a proof of $FS(0) \rightarrow FS(S(0))$. This enables us to construct a proof of $FS(S(0))$. Eventually, this process will result in the construction of a proof of Fk: in fact, it will do so after as many steps as there are Ss in the numeral k. The completability of the construction of k, by starting with 0 and repeatedly applying S, guarantees the completability of the construction of the proof of Fk, which parallels the former construction step-by-step. Mathematical induction can thus be justified intuitionistically; the proof given of Theorem 1.3 above will indeed be acceptable to the intuitionist.[19]

We have already observed that there are classically acceptable proofs that are rejected by the intuitionist. A statement can be amenable to a multitude of proofs, however, so we cannot infer from this observation alone that there are statements which the classical mathematician will assert but which the intuitionist will refrain from asserting, or which will perhaps even be refuted by the intuitionist. We will return to this question in the following chapter, where intuitionistic mathematics will be developed in greater detail. For now, it suffices to say that some statements the classical mathematician accepts as proved will, when construed intuitionistically, not be accepted as such, indeed, that they will very likely forever resist intuitionistic proof.

This must dash any residual hopes one might have entertained regarding the achievability of Frege's conception of a logicist foundation

---

[19] For further discussion, see George and Velleman (1998).

for mathematics. Earlier, we noted that Russell's Paradox signals that Frege cannot hold on to both the generality of a foundational logic and also a realist stance toward the mathematics that he wanted to found. We can now see a little better at what price realism is abandoned. Russell's Paradox suggests, then, that attempts to carry out the logicist program employing a notion of extension that is sufficiently general to qualify as logic will lead to a conception of the meaning of mathematical statements that will not underwrite the principles of inference needed to justify all of classical mathematics. The kind of foundation Frege sought is not available for the range of mathematics for which he sought it.

Intuitionists would go further, and observe that the range of mathematics that is arguably amenable to the kind of foundation Frege envisaged is not in need of such a foundation, or at least not nearly as much in need as is classical mathematics. And, in fact, intuitionists have traditionally not felt the need to offer foundations, certainly not foundations of a logical sort, for their mathematics. Recall from Chapter 1 that one bedeviling feature of mathematics is the *a priori* character of the knowledge it yields. How can we, through pure thought, discover the way things are in a domain that exists independently of us? Should a realist construal of mathematics be abandoned, however, the question must be rejected. According to the intuitionist, the mathematical world does not exist determinately and independently of us, waiting, as it were, for us to discover facts about it. It is not clear with which metaphor to replace that of discovery, but one of creation easily suggests itself. Thus we might say that, according to the intuitionist, the world of mathematics comes into being in the course of our reasoning about it. This does not mean that mathematical truths are "subjective," any more than truths about buildings are, truths that likewise only come into being as a consequence of human activity. Nor does it mean that "anything goes" in mathematics, any more than the created nature of buildings means that any structure we want will stand up. But it does suggest that the mathematical world fails to possess the kind of independence that can make it seem a mystery how *a priori* knowledge about it could ever be attained.

Another respect in which intuitionistic mathematics is less removed from the human condition, and so raises less pressing questions about how we can know or even understand it, is in its shunning of the completed infinite. The intuitionist recognizes no mathematical construction, be it a proof, a definition, or whatever, unless it is finite. Nothing can be given to one for intellectual consideration unless it is finite. This includes operations which we see can be applied *ad infinitum*: while such operations are intelligible, there is no such thing as the result of having applied one infinitely often. Human experience and thought are not only finite, of course, but also bounded in every measurable way. Consequently, intuitionism is not idealization-free. But the idealizations that it indulges in, in particular that *any* finite construction can in principle be given to us, are not ones that, from its perspective, raise the philosophical perplexity that the realist's talk of the completed infinite does.

From the intuitionist's point of view, in sum, a number of the philosophical problems for which a logicist reduction was to be the solution are to be solved by abandoning the realist perspective that creates the problems in the first place. From a classical perspective, this amounts to throwing out the baby with the bathwater: human limitations are being confusedly projected onto the world of mathematics, resulting in something that "lacks the familiarity, the convenience, the simplicity, and the beauty of our [classical] logic."[20]

What are we to make of this situation? Is the dispute without substance? Or is one position right and the other wrong? If so, is there some neutral perspective from which the dispute can be settled? Or is there no way to resolve the disagreement, the two sides offering such incommensurable ways of reasoning that there can be no noncircular rational adjudication? Or, finally, might there be a way of accommodating both perspectives, for instance by somehow acknowledging many of the intuitionist's criticisms of the classical position, while at the same time legitimating the use of classical reasoning? We shall explore this last proposal in Chapter 6. Before doing so, however, we should be more explicit about some of the features of intuitionistic mathematics.

Exercises

1. Explain why all of the inferences (24)–(27) are valid except for (25a) and (26a).

   In the following exercises, determine whether or not each inference is intuitionistically valid. (Note that they are all valid according to classical logic.)

2. (a) $\dfrac{\neg X \lor \neg Y}{\neg(X \land Y)}$; (b) $\dfrac{\neg(X \land Y)}{\neg X \lor \neg Y}$.

---

[20] Quine (1970, 87).

3. (a) $\dfrac{(X \wedge \neg Y) \to Z}{X \to (Y \vee Z)}$;   (b) $\dfrac{X \to (Y \vee Z)}{(X \wedge \neg Y) \to Z}$.

4. (a) $\dfrac{\exists x(Fx \to Gx)}{\forall x Fx \to \exists x Gx}$;   (b) $\dfrac{\forall x Fx \to \exists x Gx}{\exists x(Fx \to Gx)}$.

(Hint: For part (b), you might find it helpful to think first about why this is valid according to classical logic.)

5. (a) $\dfrac{\forall x(Fx \to Gx)}{\exists x Fx \to \exists x Gx}$;   (b) $\dfrac{\exists x Fx \to \forall x Gx}{\forall x(Fx \to Gx)}$.

6. (a) $\dfrac{\forall x Fx}{\neg \exists x \neg Fx}$;   (b) $\dfrac{\neg \exists x \neg Fx}{\forall x Fx}$;

(c) $\dfrac{\neg \exists x \neg Fx \quad \forall x(Fx \vee \neg Fx)}{\forall x Fx}$.

5

Intuitionistic Mathematics

In the last chapter, we saw that the intuitionists' view of mathematical reality leads them to interpret mathematical language differently from classical mathematicians, and as a result they do not accept all of the laws of logic that classical mathematicians accept. But the classical laws of logic are, according to the logicists, the foundation on which all of mathematics rests! If these laws are to be revised, then the work of deriving mathematics by means of these laws will also need to be revised. In this chapter, we will reexamine the development of the number systems and their properties that was carried out in Chapter 3, to see if it can be justified with intuitionistically acceptable reasoning. We will find that, while some of it can be so justified, some cannot be and will have to be either modified or discarded in intuitionistic mathematics. All proofs in this chapter will be intuitionistic proofs, unless specified otherwise.

This project was actually already started in the last chapter, where it was observed that the principle of mathematical induction is intuitionistically acceptable. In fact, much of the study of the natural numbers involves general reasoning about calculations that could, in principle, be carried out explicitly in a finite number of steps, and we have already observed that intuitionists and classical mathematicians agree that such calculations have determinate outcomes, and that they agree about what those outcomes are. As a result, most classical proofs in elementary number theory are acceptable to intuitionists.

In fact, we can make this claim more precise. It is possible to define a formal axiomatic theory based on Peano's axioms for the natural numbers, and show that many classical proofs in elementary number theory can be carried out in this formal theory. We will give a more precise definition of such a formal theory in Chapter 7, where the theory is called "PA" (for "Peano Arithmetic"). The formal theory PA uses the laws of classical logic as its inference rules, but by replacing these rules with the rules of intuitionistic logic discussed in the last chapter we can define a formal theory for intuitionistic arithmetic. The resulting theory is known as "Heyting Arithmetic" – after the Dutch mathematician Arend Heyting (1898–1980) – or "HA" for short. It is now possible to investigate the relationship between classical and intuitionistic number theory by studying the relationship between PA and HA. Such an investigation was carried out by Gödel, who proved in 1933 that every theorem of PA that does not contain either of the symbols "∨" or "∃" is also a theorem of HA.[1] (A similar theorem was proven independently by the German mathematician Gerhard Gentzen (1909–45).) Notice that in classical logic every statement is equivalent to one that does not contain "∨" or "∃", so we can conclude that for every statement P that is provable in PA there is a statement P' that is equivalent to P in classical logic (although perhaps not in intuitionistic logic) such that P' is provable in HA. We might say that every theorem of PA has a version that is acceptable to an intuitionist.

However, this is not to say that there will be no disagreements between classical mathematicians and intuitionists in the study of the natural numbers. For example, if a classical mathematician proves in PA a statement of the form $\neg \forall n P(n)$, where $P(n)$ is a predicate that does not contain either of the symbols "∨" or "∃", then by Gödel's theorem an intuitionist will also accept this theorem. But the classical mathematician will go on to claim that he has also established that $\exists n \neg P(n)$, and the intuitionist will be unwilling to accept this claim, because he does not accept the validity of rule (22b) in the last chapter. Or, for another example, a classical mathematician might prove some arithmetical statement S by showing that it is implied both by Goldbach's Conjecture and also by its negation. To the classical mathematician, Goldbach's Conjecture must be either true or false, despite our ignorance about which of these possibilities is actually the case. If each of these possibilities implies S, then S must be true. But we have already seen that intuitionists are not willing to apply the Law of the Excluded Middle to Goldbach's Conjecture. To the intuitionist, the classical mathematician's proof does not prove S; it merely tells us how we will be able to prove S once Goldbach's Conjecture has been settled. Until the conjecture is settled, we simply do not have, and do not know how to produce, a proof of S.

Nevertheless, such disagreements will be relatively rare in elementary number theory. In particular, the derivations of the basic algebraic facts about the natural numbers that we gave in Chapter 3 relied only on mathematical induction and other methods of reasoning about numerical

---

[1]   Gödel (1933).

calculation that are intuitionistically acceptable. Thus, these facts are just as acceptable to intuitionists as they are to classical mathematicians. The same can be said of the basic algebraic facts about the integers and the rational numbers.

However, matters become more complicated when we come to the real numbers. Recall that in Chapter 3 we began our study of the real numbers with the idea that a real number is picked out by a Cauchy sequence of rational numbers, where a sequence $\{a_n\}$ is called Cauchy if

(1) $\forall k \in \mathbb{Z}^+ \exists N \in \mathbb{Z}^+ \forall m, n > N(|a_m - a_n| < 1/k)$.

In intuitionistic mathematics, real numbers are also picked out by Cauchy sequences of rational numbers, but the quantification over the positive integers that appears in (1) must now be interpreted intuitionistically. For an intuitionist, it is only correct to assert (1) if we have a proof of it, and according to (9)(v) of Chapter 4, a proof of (1) must be an operation that, given a positive integer k,[2] produces a proof that there is a positive integer N with the property specified in (1). But by (9)(vi) of Chapter 4, a proof that there is a positive integer N with a certain property must consist of a specification of a particular value of N, and a proof that it has the required property. Thus, according to the intuitionist, we can assert that a sequence $\{a_n\}$ is Cauchy only if we know an operation that, given any positive integer k, will produce a positive integer N and a proof that this N is as required in (1).

For example, consider the following sequence:

(2) $a_n = \begin{cases} 1, & \text{if } n \text{ is the smallest even integer greater than 2 that} \\ & \text{cannot be written as a sum of two primes;} \\ 0, & \text{otherwise.} \end{cases}$

To a classical mathematician, this sequence is either all 0's (if Goldbach's Conjecture is true) or it is all 0's except for one 1 (if Goldbach's Conjecture is false). In either case, the sequence is easily seen to be Cauchy. But now consider the matter from the intuitionistic point of view. An intuitionist will not assert that this sequence is Cauchy unless he knows how

---

[2] For an intuitionist, one can only be given an abstract object by being given a finite description of it. For example, in the previous chapter, we assumed that natural numbers were always presented by means of numerals. Throughout this chapter, the same assumptions apply, but for simplicity's sake we will not always bother to speak explicitly of the descriptions by means of which numbers are presented.

to compute, for any given positive integer k, the positive integer N that is required in (1). For example, consider the case $k = 2$. How can we find a positive integer N such that

$$(3) \quad \forall m, n > N \left(|a_m - a_n| < \frac{1}{2} \right)?$$

If Goldbach's Conjecture is true, then any positive integer can be used for N, since for every m and n we will have

$$|a_m - a_n| = |0 - 0| = 0 < \frac{1}{2}.$$

On the other hand, if Goldbach's Conjecture is false, then we must make sure that N is at least as large as the smallest counterexample to Goldbach's Conjecture, since if this smallest counterexample were some number $n > N$ then we would have

$$|a_n - a_{n+1}| = |1 - 0| = 1 > \frac{1}{2},$$

contrary to (3). Thus, it appears that to compute N we must first settle Goldbach's Conjecture.

In fact, from an intuitionistically acceptable proof that $\{a_n\}$ is Cauchy we could produce, by means of a finite calculation, a proof of either Goldbach's Conjecture or its negation. To see how this could be done, suppose that we had a proof that $\{a_n\}$ is Cauchy. This proof would be an operation that, applied to any positive integer k, would produce the required positive integer N and a proof that this N has the required property. Applying this operation in the case $k = 2$, we could compute a number N and produce a proof of (3). By the definition of a_n, (3) implies that no number greater than N can be the smallest counterexample to Goldbach's Conjecture. But now it is just a finite calculation to check all of the numbers up to N to see if any of them is a counterexample to Goldbach's Conjecture. If, in the course of this calculation, a counterexample is found, then of course we have proven that Goldbach's Conjecture is false. If, at the end of this calculation, no counterexample has been found, then we have proven Goldbach's Conjecture.

Thus, we have an operation that, when applied to any intuitionistic proof that $\{a_n\}$ is Cauchy, will produce a proof of either Goldbach's Conjecture or its negation. Another way to say this is that we have an intuitionistic proof of the following theorem:

THEOREM 5.1.   *If the sequence* $\{a_n\}$ *defined in (2) is Cauchy, then we can assert either Goldbach's Conjecture or its negation.*

This theorem makes it very unlikely that any mathematician currently knows an intuitionistically acceptable proof that $\{a_n\}$ is Cauchy, since anyone who knows such a proof could use it to settle Goldbach's Conjecture, and no one has yet been able to settle Goldbach's Conjecture. Thus, an intuitionist would refrain from asserting that $\{a_n\}$ is Cauchy.

This is not to say that an intuitionist would assert that $\{a_n\}$ is *not* Cauchy. In fact, such an assertion would be a mistake. To see why, suppose we had a proof that $\{a_n\}$ is not Cauchy. We first claim that, using this proof, we could produce a proof that, for every positive integer t, $a_t = 0$. To see how this could be done, let t be any positive integer. By means of a finite calculation, we can determine whether or not t is the smallest counterexample to Goldbach's Conjecture, and thus we can determine whether a_t is 0 or 1. Thus, we are justified in asserting that either $a_t = 0$ or $a_t = 1$. Now, if $a_t = 1$, then according to the definition of a_n, we must have $a_n = 0$ for all $n > t$. But then it follows that the sequence is Cauchy, because for any positive integer k we can let $N = t$, and we can prove that, for any $m, n > N$, we will have

$$|a_m - a_n| = |0 - 0| = 0 < 1/k.$$

Since this contradicts our assumption that $\{a_n\}$ is not Cauchy, the possibility that $a_t = 1$ can be ruled out, so we can conclude that $a_t = 0$. Since t was arbitrary, we have proven that, for every positive integer t, $a_t = 0$, as claimed. But from this it follows easily that the sequence is Cauchy, contradicting our assumption that it is not!

Thus, the assumption that $\{a_n\}$ is not Cauchy leads to a contradiction. Does this show that the sequence is Cauchy after all? Not at all, according to the intuitionists. By deriving a contradiction from the assumption that $\{a_n\}$ is not Cauchy we have merely proven the following theorem:

THEOREM 5.2.   *It is not the case that the sequence* $\{a_n\}$ *defined in (2) is not Cauchy.*

No doubt classical mathematicians will complain at this point that the intuitionists are splitting imaginary hairs. Why not just say that the sequence is Cauchy? But to intuitionists this would be eliding an important distinction. Theorem 5.2 more accurately expresses exactly what has been proven; namely, that the assumption that the sequence is not Cauchy leads to a contradiction.

Here is another way of describing the situation. Letting P stand for the statement "the sequence $\{a_n\}$ is Cauchy," we can say that intuitionists will assert $\neg\neg P$, but they will refrain from asserting P. Thus, this example illustrates why intuitionists are not willing to use the Law of Double Negation Elimination. However, it is not a counterexample in the usual sense. A counterexample to the Law of Double Negation Elimination would be a statement P such that $\neg\neg P$ is true but P is false. However, for the statement P of our example, intuitionists would not say that P is false; they would merely refrain from asserting P, because finding grounds to justify the assertion of P is tantamount to resolving Goldbach's Conjecture, and this conjecture is still unresolved. Note that the intuitionists' mistrust of the Law of Double Negation Elimination will not be allayed if Goldbach's Conjecture is ever resolved, since we could always reformulate the example using some other unsolved problem. Brouwer used many similar examples to explain his objections to certain classical laws of logic, so such examples have come to be called *Brouwerian counterexamples*. These examples show, not that the law in question is false but, rather, that we do not have, and most likely will never have, adequate grounds to assert the law. We will see a number of other examples of Brouwerian counterexamples later in this chapter.

It is important to distinguish the example just discussed from the sequence $\{b_n\}$ defined as follows:

$$(4) \quad b_n = \begin{cases} 0, & \text{if Goldbach's Conjecture is true;} \\ 1, & \text{if Goldbach's Conjecture is false.} \end{cases}$$

Classically, this sequence is either all 0's or all 1's (depending on whether Goldbach's Conjecture is true or false), and in either case the sequence is Cauchy. But intuitionistically, (4) does not define a sequence at all, because it does not tell us how to compute any of the terms of the sequence – at least, not until Goldbach's Conjecture is resolved. To be convinced that (4) determines the value of b_n for every n we would have to be convinced that the Law of the Excluded Middle holds for Goldbach's Conjecture, and intuitionists are not convinced of this. Thus, (4) is not acceptable to intuitionists as a definition of a sequence of rational numbers.

Continuing to imitate the development of the real numbers in Chapter 3, we next define $\{a_n\} \approx \{b_n\}$ to mean

$$(5) \quad \forall k \in \mathbb{Z}^+ \exists N \in \mathbb{Z}^+ \forall n > N(|a_n - b_n| < 1/k).$$

Once again, we must be careful to interpret (5) intuitionistically. Thus, we must not assert that $\{a_n\} \approx \{b_n\}$ unless we are prepared to say how to compute, for any given $k \in \mathbb{Z}^+$, a value of N that we can prove fulfills the requirements of (5). Nevertheless, the proof that $\approx$ is an equivalence relation can be carried out exactly as in Chapter 3, and as in Chapter 3 we define real numbers to be equivalence classes of Cauchy sequences.[3]

In addition to saying how intuitionists define real numbers, it is also important to explain how they interpret quantification over the real numbers. It seems clear that, for an intuitionist, a proof of a statement of the form $\forall x \in \mathbb{R}\, Fx$ should be an operation that, given a real number x, produces a proof that F is true of x. But how can one be given a real number?

A natural number can be specified with a numeral, which is a finite specification that always succeeds in picking out a unique natural number. This is what allowed us to say, in the last chapter, that a proof of a statement of the form $\forall x \in \mathbb{N}\, Fx$ is an operation that, given a numeral n, produces a proof of Fn. Specifying real numbers is more difficult, for two reasons. First, real numbers are determined by Cauchy sequences of rational numbers, and such sequences are infinitely long. Since one cannot be given something infinite, one can never be given an entire Cauchy sequence. And yet, we have already discussed several specific Cauchy sequences in this book. How did we do it? The answer is that in each case, we gave not the sequence itself, but rather a rule by which the terms of the sequence could be determined. We will return to this issue later in this chapter, but for the moment we will only discuss Cauchy sequences that are determined by such rules.

The second difficulty in specifying real numbers is that, as we saw earlier with the sequence $\{a_n\}$ defined by (2), when a rule determining a sequence of rational numbers is given, it is not always easy to tell if the sequence is Cauchy. To be sure that a real number has been specified, we need to be given not only a rule that determines a sequence of rational numbers, but also a proof that the resulting sequence is Cauchy. This leads us to the following interpretation of quantification over the real numbers. A proof of a statement of the form $\forall x \in \mathbb{R}\, Fx$ is an operation

---

[3] We are ignoring here some complications involving the intuitionistic interpretation of the concept of an equivalence class. For more information about this issue, see the discussion of "species" in Heyting (1976, 37). For our purposes in this chapter, these complications will be unimportant. All that will matter for us is that real numbers are determined by Cauchy sequences, and that equivalent Cauchy sequences determine the same real number.

that, given a rule that determines a sequence of rational numbers and a proof that the resulting sequence is Cauchy, produces a proof that F holds of the real number determined by this sequence. Similarly, a proof of a statement of the form $\exists x \in \mathbb{R}\, Fx$ consists of a rule that determines a sequence of rational numbers, a proof that this sequence is Cauchy, and a proof that F holds of the real number determined by the sequence.

We can now define the arithmetical operations on real numbers and the ordering of the real numbers exactly as before, and begin studying their properties. Thus, if $x = [\{a_n\}]$ and $y = [\{b_n\}]$ then we have

$$x + y = [\{a_n + b_n\}],$$
$$x \cdot y = [\{a_n \cdot b_n\}],$$

and

$$x < y \text{ if and only if } \exists k \in \mathbb{Z}^+ \exists N \in \mathbb{Z}^+ \forall n > N\left(a_n + \frac{1}{k} < b_n\right).$$

The proofs that these definitions make sense are exactly as before, as are the proofs of some of the basic properties of the real numbers that follow from these definitions. For example, the classical proofs of the commutative, associative, and distributive laws are intuitionistically acceptable, as is the proof that for any real numbers x, y, and z, if $x < y$ then $x + z < y + z$.

However, there are points where intuitionists will dispute the findings of classical mathematicians. Often these disputes arise when we are working with statements that involve logical negation. For example, in some classical reasoning the following theorem is useful. As you read the classical proof, see if you can identify the step or steps at which an intuitionist would object.

CLASSICAL THEOREM 5.3.   *Suppose* $x = [\{a_n\}]$ *and* $y = [\{b_n\}]$. *If* $x \neq y$, *then*

$$\exists k \in \mathbb{Z}^+ \exists N \in \mathbb{Z}^+ \forall n > N(|a_n - b_n| > 1/k).$$

Classical Proof. Suppose $x \neq y$. Then $\{a_n\} \not\approx \{b_n\}$ or, in other words,

(6) $\neg \forall k \in \mathbb{Z}^+ \exists N \in \mathbb{Z}^+ \forall n > N(|a_n - b_n| < 1/k).$

It follows that

(7) $\exists k \in \mathbb{Z}^+ \neg \exists N \in \mathbb{Z}^+ \forall n > N(|a_n - b_n| < 1/k)$,

so we can let k_1 be some positive integer such that there is no positive integer N for which we have $\forall n > N(|a_n - b_n| < 1/k_1)$.

Since $\{a_n\}$ is Cauchy, we can choose a positive integer N_1 such that

$$(8) \quad \forall n, m > N_1\left(|a_m - a_n| < \frac{1}{3k_1}\right).$$

Similarly, since $\{b_n\}$ is Cauchy we can choose $N_2 \in \mathbb{Z}^+$ such that

$$(9) \quad \forall n, m > N_2\left(|b_m - b_n| < \frac{1}{3k_1}\right).$$

Let $k = 3k_1$, and let $N = \max(N_1, N_2)$. To complete the proof, we must show that $\forall n > N(|a_n - b_n| > 1/k)$, so let $n > N$ be arbitrary. By the choice of k_1, we know that

(10) $\neg \forall m > N(|a_m - b_m| < 1/k_1)$,

so

(11) $\exists m > N(|a_m - b_m| \geq 1/k_1)$.

Choose some $m > N$ such that $|a_m - b_m| \geq 1/k_1$. Then

$$\begin{aligned}
1/k_1 &\leq |a_m - b_m| \\
&= |a_m - a_n + a_n - b_n + b_n - b_m| \\
&\leq |a_m - a_n| + |a_n - b_n| + |b_n - b_m| \quad \text{(by the triangle inequality)} \\
&< \frac{1}{3k_1} + |a_n - b_n| + \frac{1}{3k_1} \quad \text{(by (8) and (9))}.
\end{aligned}$$

It follows that

$$|a_n - b_n| > \frac{1}{k_1} - \frac{1}{3k_1} - \frac{1}{3k_1} = \frac{1}{3k_1} = \frac{1}{k},$$

as required. ∎

There are two steps in this proof to which an intuitionist would object. Both the step from (6) to (7) and also the step from (10) to (11) involve moving from the negation of a universal statement to the existence of a counterexample to the statement. This is inference rule (22b) from Chapter 4, and we saw there that this rule is not intuitionistically valid. According to the intuitionist, the numbers k and m are introduced into this proof without justification, because no means of computing them have been given.

The intuitionistic failure of this proof motivates the following definition:

DEFINITION 5.4.   If $x = [\{a_n\}]$ and $y = [\{b_n\}]$, then we say that x *lies apart from* y, and write $x \# y$, if

$$\exists k \in \mathbb{Z}^+ \exists N \in \mathbb{Z}^+ \forall n > N(|a_n - b_n| > 1/k).$$

As usual, to justify this definition one must prove that it is unambiguous by showing that a different choice of Cauchy sequences representing x and y would give the same results (see exercise 1). Intuitively, $x \# y$ means not merely that x and y are different, but that there is a positive separation between them. It is not hard to prove that $x \# y$ implies $x \neq y$ (see exercise 2), but for intuitionists there is no reason to believe that $x \neq y$ implies $x \# y$. For classical mathematicians, Classical Theorem 5.3 shows that $x \neq y$ implies $x \# y$, and hence that $x \neq y$ and $x \# y$ are equivalent.

Although they are unconvinced by the classical mathematician's proof of Classical Theorem 5.3, intuitionists can prove the following theorem, which to classical mathematicians is equivalent to Classical Theorem 5.3:

THEOREM 5.5.   *If it is not the case that $x \# y$, then $x = y$.*

Proof. Suppose that $x = [\{a_n\}]$ and $y = [\{b_n\}]$. We assume that it is not the case that $x \# y$. To show that $x = y$, we must prove that

$$\forall k \in \mathbb{Z}^+ \exists N \in \mathbb{Z}^+ \forall n > N(|a_n - b_n| < 1/k),$$

so let k be an arbitrary positive integer. Since $\{a_n\}$ and $\{b_n\}$ are Cauchy sequences, we can find positive integers N_1 and N_2 such that

$$(12) \quad \forall m, n > N_1 \left(|a_m - a_n| < \frac{1}{3k} \right)$$

and

(13) $\forall m, n > N_2 \left(|b_m - b_n| < \dfrac{1}{3k} \right).$

Let $N = \max(N_1, N_2)$. We must now show that $\forall n > N(|a_n - b_n| < 1/k)$, so our next step is to let $n > N$ be arbitrary.

Note that a_n and b_n are rational numbers, so we can, by means of a finite calculation, compute $|a_n - b_n|$ and compare it to $1/k$. Thus we are justified in asserting that either $|a_n - b_n| < 1/k$ or $|a_n - b_n| \geq 1/k$. To complete the proof, it suffices to rule out the second of these two possibilities, so suppose that $|a_n - b_n| \geq 1/k$. Then, for all $m > N$, we have

$$
\begin{aligned}
\frac{1}{k} &\leq |a_n - b_n| \\
&= |a_n - a_m + a_m - b_m + b_m - b_n| \\
&\leq |a_n - a_m| + |a_m - b_m| + |b_m - b_n| \quad &\text{(by the triangle inequality)} \\
&< \frac{1}{3k} + |a_m - b_m| + \frac{1}{3k} \quad &\text{(by (12) and (13)).}
\end{aligned}
$$

Thus

$$
|a_m - b_m| > \frac{1}{k} - \frac{1}{3k} - \frac{1}{3k} = \frac{1}{3k}.
$$

Since m was arbitrary, we have shown that $\forall m > N(|a_m - b_m| > 1/(3k))$. But according to Definition 5.4, this means that $x \mathbin{\#} y$, contrary to our original assumption. Therefore we can rule out the possibility that $|a_n - b_n| \geq 1/k$, and so we must have $|a_n - b_n| < 1/k$, as required. ∎

COROLLARY 5.6.   *If it is not the case that $x \neq y$, then $x = y$.*

Proof. Assume that it is not the case that $x \neq y$. If $x \mathbin{\#} y$ then $x \neq y$, contradicting this assumption. Therefore it cannot be the case that $x \mathbin{\#} y$, so by Theorem 5.5, $x = y$. ∎

Corollary 5.6 shows that the Law of Double Negation Elimination can hold for some statements. When we say that intuitionists reject this law, we do not mean that they believe that it never holds. Rather, we mean that they are not willing to assume the law; it must be proven in any case in which it is to be used.

Another source of disagreements between classical mathematicians and intuitionists is the use of the word "or" in mathematical statements. For example, consider the classical theorem that, for any real numbers x

and y, if $x \cdot y = 0$ then either $x = 0$ or $y = 0$. Here is a Brouwerian counterexample to this theorem. We define sequences $\{a_n\}$ and $\{b_n\}$ as follows. For any positive integer n, if there is some $j \leq n$ such that a sequence of 100 consecutive 9's appears in the decimal expansion of π for the first time starting at the jth digit, then if j is even we let $a_n = 1/j$ and $b_n = 1/n$, and if j is odd we let $a_n = 1/n$ and $b_n = 1/j$. If there is no such j, then we let $a_n = b_n = 1/n$. For example, if a sequence of 100 consecutive 9's appears in the decimal expansion of π for the first time starting at digit number 571, then the sequences $\{a_n\}$ and $\{b_n\}$ are as follows:

$$\{a_n\} : 1, \frac{1}{2}, \frac{1}{3}, \frac{1}{4}, \ldots, \frac{1}{570}, \frac{1}{571}, \frac{1}{572}, \frac{1}{573}, \cdots$$

$$\{b_n\} : 1, \frac{1}{2}, \frac{1}{3}, \frac{1}{4}, \ldots, \frac{1}{570}, \frac{1}{571}, \frac{1}{571}, \frac{1}{571}, \cdots$$

Note that in this case $[\{a_n\}] = 0$ and $[\{b_n\}] = 1/571 > 0$. Similarly, if a sequence of 100 consecutive 9's first appears at an even numbered digit, then $[\{a_n\}] > 0$ and $[\{b_n\}] = 0$. If there is no sequence of 100 consecutive 9's in the decimal expansion of π, then $[\{a_n\}] = [\{b_n\}] = 0$.

Even without knowing whether or not there is a sequence of 100 consecutive 9's in the decimal expansion of π, we can prove that $\{a_n\}$ and $\{b_n\}$ are Cauchy sequences. To see why, let k be an arbitrary positive integer, and let $N = k$. We claim that $\forall m, n > N(|a_m - a_n| < 1/k)$. To prove this, first note that by computing the first $N + 100$ digits in the decimal expansion of π (a finite calculation), we can determine whether or not a sequence of 100 consecutive 9's appears in the decimal expansion of π for the first time starting at the jth digit for any even integer $j \leq N$. If so, then for all $m, n > N$ we have

$$|a_m - a_n| = \left| \frac{1}{j} - \frac{1}{j} \right| = 0 < \frac{1}{k}.$$

If not, then further calculation would be required to find the exact value of a_n for any $n > N$, but we can tell from the definition of a_n, without doing the calculation, that $0 < a_n < 1/N$. It follows that, for any $m, n > N$,

$$|a_m - a_n| < \frac{1}{N} = \frac{1}{k}.$$

A similar argument shows that $\{b_n\}$ is also Cauchy.

Since $\{a_n\}$ and $\{b_n\}$ are Cauchy, we can let $r = [\{a_n\}]$ and $s = [\{b_n\}]$. According to the definitions of a_n and b_n, for every n, either $a_n \cdot b_n = 1/n^2$ or $a_n \cdot b_n = 1/(j \cdot n)$ for some positive integer j, and in both cases we have $0 < a_n \cdot b_n < 1/n$. It follows that $r \cdot s = [\{a_n \cdot b_n\}] = 0$. But to give an intuitionistically acceptable proof that either $r = 0$ or $s = 0$, we would have to be able to say which of these statements is true, and prove it. By reasoning given earlier, a proof that $r = 0$ would show that a sequence of 100 consecutive 9's does not appear in the decimal expansion of π for the first time at an even numbered digit, and a proof that $s = 0$ would show that such a sequence does not first appear at an odd numbered digit. But it is not currently known whether or not there is a sequence of 100 consecutive 9's in the decimal expansion of π, and if so whether the first such sequence starts at an even or an odd numbered digit. Thus, no intuitionistically acceptable proof that either $r = 0$ or $s = 0$ is known.

Thus, an intuitionist would be unwilling to assert the theorem that, for all real numbers x and y, if $x \cdot y = 0$ then either $x = 0$ or $y = 0$. But the intuitionist can prove the following theorem:

THEOREM 5.7.   *If $x \# 0$ and $y \# 0$, then $x \cdot y \# 0$.*

Proof. Suppose $x = [\{a_n\}]$ and $y = [\{b_n\}]$. Since $x \# 0$ and $y \# 0$, we can find positive integers k_1, N_1, k_2, and N_2 such that

$$\forall n > N_1 \left(|a_n| > \frac{1}{k_1} \right)$$

and

$$\forall n > N_2 \left(|b_n| > \frac{1}{k_2} \right).$$

Let $N = \max(N_1, N_2)$ and let $k = k_1 \cdot k_2$. Then, for every $n > N$,

$$|a_n \cdot b_n| = |a_n| \cdot |b_n| > \frac{1}{k_1} \cdot \frac{1}{k_2} = \frac{1}{k}.$$

Therefore $x \cdot y = [\{a_n \cdot b_n\}] \# 0$.   ■

COROLLARY 5.8.   *If $x \# 0$ and $x \cdot y = 0$, then $y = 0$.*

Proof. Suppose $x \# 0$ and $x \cdot y = 0$. If $y \# 0$ then, by Theorem 5.7, $x \cdot y \# 0$, so $x \cdot y \neq 0$. But we have assumed $x \cdot y = 0$, so this is a contradiction. Therefore it is not the case that $y \# 0$, so by Theorem 5.5, $y = 0$. ∎

COROLLARY 5.9.   *If $x \cdot y = 0$, then it is not the case that $x \neq 0$ and $y \neq 0$.*

Proof. Suppose $x \cdot y = 0$, and suppose also that $x \neq 0$ and $y \neq 0$. If $x \# 0$ then, by Corollary 5.8, $y = 0$, contradicting our assumption that $y \neq 0$. Therefore it is not the case that $x \# 0$, and so by Theorem 5.5, $x = 0$. But this contradicts our assumption that $x \neq 0$. ∎

We have already given an example of real numbers r and s such that we cannot assert that either $r = 0$ or $s = 0$, although we can assert that $r \cdot s = 0$. By Corollary 5.9, we can also assert that it is not the case that $r \neq 0$ and $s \neq 0$. Thus, we have a Brouwerian counterexample to an instance of one of DeMorgan's Laws.

Note that if a sequence of 100 consecutive 9's appears in the decimal expansion of π for the first time starting at an odd numbered digit, then $r < s$; if such a sequence first appears starting at an even numbered digit, then $s < r$; and if there is no such sequence, then $r = s$. Since the digits of π problem is unsolved, at this time we can neither assert nor deny any of the statements $r < s$, $s < r$, or $r = s$. Thus these numbers also serve as a Brouwerian counterexample to the trichotomy law. But we can prove the following theorem:

THEOREM 5.10.   *If $x \# y$, then either $x < y$ or $x > y$.*

Proof. Suppose that $x = [\{a_n\}]$ and $y = [\{b_n\}]$, and assume that $x \# y$. Then, by the definition of "lies apart," we can find positive integers k_1 and N_1 such that

(14) $\forall n > N_1(|a_n - b_n| > 1/k_1)$.

Let $k = 3k_1$. Since $\{a_n\}$ and $\{b_n\}$ are Cauchy, we can find positive integers N_2 and N_3 such that

(15) $\forall m, n > N_2(|a_m - a_n| < 1/k)$

and

(16) $\forall m, n > N_3(|b_m - b_n| < 1/k)$.

Let $N = \max(N_1, N_2, N_3)$ and let $m = N + 1$. By (14), $|a_m - b_m| > 1/k_1$, so either $a_m < b_m - 1/k_1$ or $b_m < a_m - 1/k_1$. (Note that a_m and b_m are rational numbers, so determining which of these possibilities is the case can be done with a finite calculation.)

Case 1: $a_m < b_m - 1/k_1$. In this case, we claim $\forall n > N(a_n + 1/k < b_n)$, from which it follows that $x < y$. To see why, let $n > N$ be arbitrary. By (15) and (16), $a_n < a_m + 1/k$ and $b_m < b_n + 1/k$. Therefore

$$a_n + \frac{1}{k} < a_m + \frac{1}{k} + \frac{1}{k} = a_m + \frac{2}{3k_1} < b_m - \frac{1}{k_1} + \frac{2}{3k_1}$$

$$= b_m - \frac{1}{3k_1} = b_m - \frac{1}{k} < b_n + \frac{1}{k} - \frac{1}{k} = b_n.$$

Case 2: $b_m < a_m - 1/k_1$. In this case, a similar argument shows that $y < x$. ∎

COROLLARY 5.11. *If either $x \,\#\, y$ or it is not the case that $x \,\#\, y$, then either $x < y$, $y < x$, or $x = y$.*

Proof. Combine Theorems 5.5 and 5.10. ∎

Consider once again our real numbers r and s, for which we cannot assert that either $r < s$, $s < r$, or $r = s$. By Corollary 5.11, we also cannot assert that either $r \,\#\, s$ or it is not the case that $r \,\#\, s$. Thus, we have a Brouwerian counterexample to another instance of the Law of the Excluded Middle.

So far we have considered only Cauchy sequences that are specified by rules that determine how the terms are computed. But an examination of the proofs of the theorems we have given shows that these rules have played no role in our proofs. Our proofs have discussed the terms of Cauchy sequences, and the values of N and k that arise when working with expressions such as "Cauchy" or "lies apart," but in no case have we drawn conclusions by reasoning about the rules by which the terms of our Cauchy sequences are determined. This suggests the possibility of recognizing a broader range of applicability for the theorems we have proven.

And there is reason to seek such a broadening of the range of applicability of our theorems. To classical mathematicians, the real numbers that are represented by rule-determined Cauchy sequences make up only a very small part of the number line. For example, there are only countably many expressions in the English language that could specify a rule for determining a Cauchy sequence, but according to Theorem 3.45 the set of all real numbers is uncountable. The classical mathematicians' response to this observation is to insist that the number line includes real numbers determined by *all possible* Cauchy sequences, and these possibilities are not exhausted by those sequences that are determined by rules. For example, a sequence might be generated by some process that includes a random element, such as flipping a coin. Of course, classical mathematicians regard the collection of *all possible* Cauchy sequences as a determinate totality, a view that to the intuitionists is incoherent. But if intuitionists want to do justice to the conception of the number line that underlies the classical mathematicians' understanding of the real numbers, then it would be desirable to be able to apply intuitionistic reasoning to Cauchy sequences that are not rule-determined. Brouwer therefore came to believe that the intuitionistic theory of the real numbers should allow for Cauchy sequences generated by an infinite sequence of free choices of rational numbers. Such sequences are called *choice sequences*, or *infinitely proceeding sequences*.

It might seem at first that any discussion of choice sequences, which require infinitely many choices for their construction, would violate the intuitionists' insistence that there is no completed infinite. But this would be the case only if we treated choice sequences as completed objects. We can maintain the intuitionists' belief that all infinity is potential if we instead treat choice sequences as sequences that can be continued indefinitely but can never be completed. We can list finitely many terms of a choice sequence, and any such finite listing can be extended by listing further terms, but we can never list all of the terms of the sequence. All reasoning about choice sequences must recognize their fundamental incompletability.

It might help to think of a choice sequence as a sequence of rational numbers that is generated by *someone else*. We can ask the sequence-generator for as many terms of the sequence as we want, and he will supply them. To guarantee that the sequence is Cauchy, we must also imagine that we can ask, for any positive integer k, that the sequence-generator specify a positive integer N and promise that all terms beyond term number N will differ from each other by less than $1/k$. In the course of generating terms of the sequence, the sequence-generator may also, if he wishes, make other promises about terms to be generated in the future. We do not know how the sequence-generator is producing the terms of the sequence; he might be using a rule to determine the terms of the sequence, or he might be rolling a die or flipping a coin. All we know is that he will produce as many terms as we want, and that he will make and live up to those promises necessary to guarantee that the sequence is

Cauchy. The study of choice sequences can be thought of as the study of the behavior of such sequence-generators.

All of the definitions and theorems about real numbers that we have discussed can be applied to real numbers that are determined by choice sequences. For example, consider our definition of addition of real numbers. If $x = [\{a_n\}]$ and $y = [\{b_n\}]$, then we have defined $x + y$ to be $[\{a_n + b_n\}]$. This definition makes perfectly good sense even if the sequences $\{a_n\}$ and $\{b_n\}$ are choice sequences. If there is no rule that determines the values of a_n and b_n, then there may also be no rule that determines $a_n + b_n$. Nevertheless, if we can somehow be supplied with as many terms of $\{a_n\}$ and $\{b_n\}$ as we want, then, by adding these terms, we can produce as many terms of the sequence $\{a_n + b_n\}$ as we might want. Furthermore, if the production of the sequences $\{a_n\}$ and $\{b_n\}$ lives up to the promises necessary to guarantee that these sequences are Cauchy, then it can be shown that the sequence $\{a_n + b_n\}$ will also live up to promises guaranteeing that it is Cauchy.

Let us consider one more example. In Theorem 5.10, the real numbers x and y might be specified by choice sequences. What the theorem tells us is that if we are supplied with promises about the terms of these sequences that are sufficient to guarantee that $x \# y$, then we will be able to make predictions about the terms of the sequences that are sufficient to justify either the conclusion that $x < y$ or the conclusion that $y < x$. As long as the terms of the sequences, as they are produced, continue to live up to the promises that have been made, we can be sure that our predictions will be fulfilled.

Consider a choice sequence that has been partially determined by specifying a finite number of terms of the sequence, and perhaps also by making promises about the terms that are yet to be specified. The range of possibilities for how the sequence might be continued can be thought of as forming an infinite tree, with different branches of the tree corresponding to different ways in which the sequence might be continued. Each way of completing the sequence corresponds to a real number, and so the whole tree can be thought of as representing a set of real numbers. Such tree-representations of sets of real numbers are called *spreads*. For example, it is not hard to show that every interval on the number line can be represented as a spread. We will not pursue the subject of choice sequences and spreads further here.[4]

---

[4] We have given only the idea behind spreads; the precise definition is somewhat more complicated. The reader interested in learning more about choice sequences and spreads should consult Heyting (1976), Dummett (2000), or Troelstra and van Dalen (1988).

Some further interesting differences between intuitionistic and classical mathematics will emerge if we study functions from the real numbers to the real numbers. For example, recall that in Chapter 3 we proved the Intermediate Value Theorem, which says that if f is a continuous function from $\mathbb{R}$ to $\mathbb{R}$, $u < v$, and $f(u) < m < f(v)$, then there is a number w such that $u < w < v$ and $f(w) = m$. In other words, if a continuous curve in the plane is below the horizontal line $y = m$ at one point and above it at another, then it must cross the line somewhere in between. It is hard to imagine how a continuous curve could get from one side of a line to the other without crossing it. However, despite its plausibility, the Intermediate Value Theorem is not intuitionistically acceptable. In fact, we can give a Brouwerian counterexample to it. Let r and s be the real numbers discussed earlier for which we can neither assert nor deny any of the statements $r < s, s < r$, or $r = s$. Let $t = r - s$. Then we can neither assert nor deny any of the statements $t < 0, t > 0$, or $t = 0$. We now define a function f for which we cannot assert the Intermediate Value Theorem. To a classical mathematician, the function we have in mind could be defined by the formula

$$(17) \quad f(x) = \begin{cases} t + x - 1, & \text{if } x < 1; \\ t, & \text{if } 1 \le x \le 2; \\ t + x - 2, & \text{if } x > 2. \end{cases}$$

However, this definition would not be acceptable to an intuitionist, because one must use the trichotomy law to be sure that (17) succeeds in defining $f(x)$ for all values of x, and trichotomy is not intuitionistically acceptable. Equation (17) does not tell us how to compute $f(x)$ if x is a number for which we cannot assert either $x < 1$, $1 \le x \le 2$, or $x > 2$.

To give an intuitionistically acceptable definition of f, we proceed slightly differently. Let $\{c_n\}$ be a Cauchy sequence of rational numbers such that $t = [\{c_n\}]$. Suppose $x = [\{a_n\}]$ is any real number. To compute $f(x)$, we first define a sequence $\{b_n\}$ as follows:

$$(18) \quad b_n = \begin{cases} c_n + a_n - 1, & \text{if } a_n < 1; \\ c_n, & \text{if } 1 \le a_n \le 2; \\ c_n + a_n - 2, & \text{if } a_n > 2. \end{cases}$$

Note that a_n is a rational number and that trichotomy holds for rational numbers; so (18) succeeds in defining b_n for every n. Let $f(x) = [\{b_n\}]$. We leave it as an exercise for the reader to verify that $\{b_n\}$ is Cauchy (see exercise 8), so our definition of $f(x)$ makes sense, and that a choice of a different Cauchy sequence representing x would lead to the same answer for $f(x)$ (see exercise 9), so our definition is unambiguous.

It follows from (18) that (17) does give the correct values of $f(x)$ for those real numbers x for which we can assert either $x < 1$, $1 \leq x \leq 2$, or $x > 2$. The graph of f in each of the three cases $t < 0$, $t = 0$, and $t > 0$ is shown in Figure 5.1. Note that to a classical mathematician one of these three cases represents the truth about f, although we do not know which, but to an intuitionist it would be unacceptable to describe the situation in this way. It is not hard to show that f is continuous, $f(0) = t - 1 < 0$, and $f(3) = t + 1 > 0$. Thus, if we could prove the Intermediate Value Theorem, then we could prove that there is a number w such that $0 < w < 3$ and $f(w) = 0$.

Of course, for such a proof to be intuitionistically acceptable it would have to tell us how to compute the terms in a Cauchy sequence $\{a_n\}$ such that if $w = [\{a_n\}]$, then $f(w) = 0$. For $\{a_n\}$ to be Cauchy, we must be able to find a number N such that

$$(19) \quad \forall m, n > N\left(|a_m - a_n| < \frac{1}{4}\right).$$

Let $m = N + 1$. By a finite calculation we can determine whether $a_m \leq 3/2$ or $a_m > 3/2$. Suppose first that $a_m \leq 3/2$. Then, by (19), for all $n > N$ we have $|a_m - a_n| < 1/4$, and therefore $a_n < 3/2 + 1/4 = 7/4$. Thus $\forall n > N(a_n + 1/4 < 2)$, so $w = [\{a_n\}] < 2$. But it is clear from Figure 5.1 that if $t < 0$ then $w > 2$, so we can conclude that it is not the case that $t < 0$. Similar reasoning shows that if $a_m > 3/2$, then it is not the case that $t > 0$.

Thus, from an intuitionistically acceptable proof of the Intermediate Value Theorem we would be able to produce a proof that either $t \not< 0$ or $t \not> 0$. But we do not know how to produce such a proof, so we must not have an intuitionistically acceptable proof of the Intermediate Value Theorem.

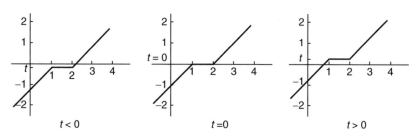

Figure 5.1

As we have seen in several previous examples, often when a classical theorem has no intuitionistic proof, there is a related theorem that is intuitionistically provable. In the case of the Intermediate Value Theorem, we will give two related intuitionistic theorems. The first is motivated by the fact that our Brouwerian counterexample was a function f with the property that $f(x)$ was very close to 0 for all x between 1 and 2. If we add a hypothesis to the theorem that rules out this kind of situation, then the theorem can be proven intuitionistically:

THEOREM 5.12.   *Suppose f is a continuous function from $\mathbb{R}$ to $\mathbb{R}$, $u < v$, $f(u) < m < f(v)$, and, for all numbers p and q, if $u \leq p < q \leq v$ then there is a number r such that $p < r < q$ and $f(r) \# m$. Then there is a number w such that $u < w < v$ and $f(w) = m$.*

Sketch of Proof. It will be convenient to work with rational numbers in our proof, so we begin by replacing u and v with nearby rational numbers. Since f is continuous at u and v and $f(u) < m < f(v)$, we can find rational numbers a_0 and b_0 such that $u \leq a_0 < b_0 \leq v$, and $f(a_0) < m < f(b_0)$. Now divide the interval from a_0 to b_0 into three equal pieces at points p and q, with $a_0 < p < q < b_0$. By hypothesis, there is a number r such that $p < r < q$ and $f(r) \# m$. Since f is continuous at r, we can find a nearby rational number c_0 such that $p < c_0 < q$ and $f(c_0) \# m$. By Theorem 5.10, either $f(c_0) < m$ or $f(c_0) > m$. If $f(c_0) < m$, then let $a_1 = c_0$ and let $b_1 = b_0$; if $f(c_0) > m$, then let $a_1 = a_0$ and $b_1 = c_0$. In both cases, we have $f(a_1) < m < f(b_1)$, $a_0 \leq a_1 < b_1 \leq b_0$, $b_1 - a_1 < 2(b_0 - a_0)/3$, and a_1 and b_1 are rational. By repeating this process, we can generate sequences $\{a_n\}$ and $\{b_n\}$ of rational numbers such that, for every n, $f(a_n) < m < f(b_n)$, $a_n \leq a_{n+1} < b_{n+1} \leq b_n$, and $b_{n+1} - a_{n+1} < 2(b_n - a_n)/3$. It is not hard to show that these sequences are Cauchy and $\{a_n\} \approx \{b_n\}$, so we can let $w = [\{a_n\}] = [\{b_n\}]$. Using the fact that f is continuous at w and $f(a_n) < m < f(b_n)$, it can then be shown that $f(w) = m$ (see exercise 10).   ∎

In Chapter 3, we applied the Intermediate Value Theorem to the function $f(x) = x^2$ to show that there is a real number w such that $w^2 = 2$. It is not hard to show that this function satisfies the additional hypothesis in Theorem 5.12, and therefore we have an intuitionistically acceptable proof of the existence of $\sqrt{2}$.

If we do not add any additional hypotheses to the theorem, then we must settle for a weaker conclusion if we are to make the theorem acceptable to intuitionists. In the following theorem, the conclusion has been weakened to say, not that $f(w) = m$, but merely that $f(w)$ is very close to m.

THEOREM 5.13. *Suppose f is a continuous function from $\mathbb{R}$ to $\mathbb{R}$, $u < v$, $f(u) < m < f(v)$, and $k \in \mathbb{Z}^+$. Then there is a number w such that $u < w < v$ and $|f(w) - m| < 1/k$.*

Sketch of Proof. The proof is very similar to the proof of Theorem 5.12. We construct Cauchy sequences $\{a_n\}$ and $\{b_n\}$ such that, for all n, $a_n \leq a_{n+1} < b_{n+1} \leq b_n$, $b_{n+1} - a_{n+1} = (b_n - a_n)/2$, $f(a_n) < m + 1/(2k)$, and $f(b_n) > m - 1/(2k)$, and then let $w = [\{a_n\}] = [\{b_n\}]$ (see exercise 11). ∎

The Intermediate Value Theorem is concerned only with continuous functions, but classical mathematicians also study functions that are not continuous. For example, a classical mathematician might study the function g defined by the equation

$$(20) \quad g(x) = \begin{cases} 0, & \text{if } x \leq 0; \\ 1, & \text{if } x > 0. \end{cases}$$

This function is not continuous, because its graph makes a sudden jump at $x = 0$. Of course, to an intuitionist, (20) is unacceptable as a definition of a function for the same reason that (17) was unacceptable. Equation (20) tells us how to compute $g(x)$ only for those values of x for which we can assert either $x \leq 0$ or $x > 0$. It does not tell us, for example, how to compute $g(t)$, where t is the number we used in defining our Brouwerian counterexample to the Intermediate Value Theorem.

In the case of (17), we were able to resolve the problem by working with Cauchy sequences that determine x and $f(x)$, and using (18) to give the relationship between these sequences. Let us see if we can do something similar with (20). It is tempting to proceed as follows. Suppose $x = [\{a_n\}]$, and let $\{b_n\}$ be the sequence defined by the equation

$$(21) \quad b_n = \begin{cases} 0, & \text{if } a_n \leq 0; \\ 1, & \text{if } a_n > 0. \end{cases}$$

Can we now define $g(x)$ to be $[\{b_n\}]$? The answer is no, because $\{b_n\}$ might not be a Cauchy sequence. For example, suppose $\{a_n\}$ is the sequence $1, -1/2, 1/3, -1/4, \ldots$ Then $\{a_n\}$ is Cauchy and $[\{a_n\}] = 0$, but $\{b_n\}$ is the sequence $1, 0, 1, 0, \ldots$, which is not Cauchy.

In fact, similar problems will plague any attempt at giving an intuitionistically acceptable definition of a discontinuous function f. To see why, suppose $x = [\{a_n\}]$, and we are trying to say how to compute the terms of a Cauchy sequence $\{b_n\}$ such that $f(x) = [\{b_n\}]$. Since $\{a_n\}$ might be a choice sequence, there might be no rule that determines the value of a_n for every n, and we can never have a list of all of the infinitely many terms of the sequence. In the course of any calculation, the only information we can obtain about $\{a_n\}$ is a list of finitely many terms, and a list of finitely many promises about the remaining terms of the kind that are needed to guarantee that the sequence is Cauchy. Thus, the calculation of any particular term b_n can only make use of such a finite list of information about $\{a_n\}$. This finite list does not determine the value of x precisely, but only narrows it down to a small range of possible values – say, from $x - 1/j$ to $x + 1/j$ for some positive integer j. (We might think of this range of possible values as being represented by a spread.) It follows that, for any n, the value of b_n that we compute would be the same for certain other Cauchy sequences $\{a'_n\}$, and there must be some positive integer j such that the real numbers represented by these other Cauchy sequences would include all real numbers between $x - 1/j$ and $x + 1/j$. Similar considerations apply to the proof that $\{b_n\}$ is Cauchy. For every positive integer k, we must be able to compute a positive integer N such that $\forall m, n > N(|b_m - b_n| < 1/k)$, but the value of N we compute must be the same for certain other Cauchy sequences representing all real numbers in the range from $x - 1/j$ to $x + 1/j$, for some j.

It follows from these conclusions that f must be continuous at x. Recall that, according to Definition 3.42, this means that

$$(22)\quad \forall k \in \mathbb{Z}^+ \exists j \in \mathbb{Z}^+ \forall w \in \mathbb{R}(|x - w| < 1/j \rightarrow |f(x) - f(w)| < 1/k).$$

To prove this, let k is a positive integer. Since $\{b_n\}$ is Cauchy, we can compute a positive integer N such that

$$(23)\quad \forall m, n > N\left(|b_m - b_n| < \frac{1}{3k}\right),$$

and we can also then compute b_{N+1}. Note that for all $n > N$ we have

$$b_{N+1} - \frac{1}{3k} < b_n < b_{N+1} + \frac{1}{3k},$$

from which it follows that

(24) $b_{N+1} - \dfrac{1}{3k} \le f(x) \le b_{N+1} + \dfrac{1}{3k}.$

By the arguments in the preceding paragraph, the computations of N and b_{N+1} use only a finite amount of information about $\{a_n\}$, and therefore there is a positive integer j such that the same values of N and b_{N+1} would be computed for certain Cauchy sequences representing all real numbers between $x - 1/j$ and $x + 1/j$. Thus, using the same reasoning that was used to prove (24), we can conclude that for all w between $x - 1/j$ and $x + 1/j$,

(25) $b_{N+1} - \dfrac{1}{3k} \le f(w) \le b_{N+1} + \dfrac{1}{3k}.$

But now combining (24) and (25) we can say that

(26) $\forall w \in \mathbb{R}(|x - w| < 1/j \rightarrow |f(x) - f(w)| < 1/k),$

as required in the definition of continuity.

   Although this reasoning is admittedly somewhat sketchy, we hope it will at least make plausible the following intuitionistic theorem.

THEOREM 5.14.   *If f is a function from* $\mathbb{R}$ *to* $\mathbb{R}$, *then f is continuous.*

   We have seen in this chapter that it is possible to do mathematics using intuitionistic methods, but the results are rather different from those in classical mathematics. In particular, in calculus, which is concerned primarily with properties of functions from $\mathbb{R}$ to $\mathbb{R}$, there are substantial differences between the conclusions of intuitionists and classical mathematicians. Classical mathematicians often work with discontinuous functions, but, in the light of Theorem 5.14, intuitionists do not even recognize the existence of such functions. And even in the study of continuous functions, some theorems that are fundamental to classical mathematics, such as the Intermediate Value Theorem, are not accepted by intuitionists. To an intuitionist, these differences represent corrections of the confusions inherent in the classical approach to mathematics. The intuitionist is not bothered by the fact that some theorems considered important by classical mathematicians must be rejected; these theorems were the result of incoherent reasoning, and their rejection is therefore not a loss but, rather, a step forward in our mathematical understanding. But to a classical mathematician, these differences represent a failure of the intuitionistic methods to yield many of the most important truths of mathematics. As the German mathematician Hermann Weyl (1885–1955) put it, "With Brouwer, mathematics gains the highest intuitive clarity; his doctrine is idealism in mathematics thought to the end. But, full of pain, the mathematician sees the greatest part of his towering theories dissolve in fog."[5]

Exercises

Note: In all exercises, all proofs should be written using only intuitionistically valid reasoning.

1. Prove that if $\{a_n\} \approx \{a'_n\}$, $\{b_n\} \approx \{b'_n\}$, and $\exists k \in \mathbb{Z}^+ \exists N \in \mathbb{Z}^+$ $\forall n > N(|a_n - b_n| > 1/k)$, then $\exists k \in \mathbb{Z}^+ \exists N \in \mathbb{Z}^+ \forall n > N(|a'_n - b'_n| > 1/k)$. (Note: This justifies Definition 5.4.)

2. Complete the following proof that if $x \# y$ then $x \neq y$:

   Suppose $x = [\{a_n\}]$ and $y = [\{b_n\}]$, where $\{a_n\}$ and $\{b_n\}$ are Cauchy sequences. Suppose $x \# y$. Then

   $$\exists k \in \mathbb{Z}^+ \exists N \in \mathbb{Z}^+ \forall n > N\left(|a_n - b_n| > \frac{1}{k}\right).$$

   To prove that $x \neq y$, we must show that the assumption that $x = y$ leads to a contradiction, so suppose that $x = y$. Then $\{a_n\} \approx \{b_n\}$ or, in other words,

   $$\forall k \in \mathbb{Z}^+ \exists N \in \mathbb{Z}^+ \forall n > N\left(|a_n - b_n| < \frac{1}{k}\right).$$

   (Now you finish the proof by deriving a contradiction.)

3. Complete the following proof that $x \# y \rightarrow \forall z(x \# z \lor y \# z)$:

   Suppose $x = [\{a_n\}]$, $y = [\{b_n\}]$, and $z = [\{c_n\}]$, where $\{a_n\}$, $\{b_n\}$, and $\{c_n\}$ are Cauchy sequences. Suppose $x \# y$. Then by definition of # we can choose k_1 and N_1 such that

   $$\forall n > N_1(|a_n - b_n| > 1/k_1).$$

   Since $\{a_n\}$, $\{b_n\}$, and $\{c_n\}$ are Cauchy, we can choose N_2, N_3, and N_4 such that

   $$\forall n, m > N_2\left(|a_n - a_m| < \frac{1}{8k_1}\right),$$

---

[5] Weyl (1925–7). The quotation is from p. 136.

$$\forall n, m > N_3 \left(|b_n - b_m| < \frac{1}{8k_1} \right),$$

and

$$\forall n, m > N_4 \left(|c_n - c_m| < \frac{1}{8k_1} \right).$$

Let $N = \max(N_1, N_2, N_3, N_4)$, and let $m = N + 1$. Then $m > N_1$, so by the choice of N_1, $|a_m - b_m| > 1/k_1$.

CLAIM. Either $|a_m - c_m| > \dfrac{1}{2k_1}$ or $|b_m - c_m| > \dfrac{1}{2k_1}$.

Proof of Claim. The key point is that, since all numbers involved are *rational*, we can, by means of a finite calculation, determine whether or not either of these inequalities holds. Thus, to prove the claim it suffices to assume that $|a_m - c_m| \leq 1/(2k_1)$ and $|b_m - c_m| \leq 1/(2k_1)$, and then derive a contradiction.

(Now you finish the proof of the claim by deriving the required contradiction, and then finish the proof of the theorem by showing that if $|a_m - c_m| > 1/(2k_1)$ then $x \# z$, and if $|b_m - c_m| > 1/(2k_1)$ then $y \# z$.)

4. Prove that if $x < y$ and $y < z$, then $x < z$.
5. Prove that if either $x < y$ or $y < x$, then $x \# y$. (This is the converse of Theorem 5.10.)
6. Prove that if $x \not< y$ and $y \not< x$, then $x = y$. (Hint: Use Theorems 5.5 and 5.10.)
7. Prove that if $x < y$, then $\forall z (z < y \lor x < z)$. (Hint: This is similar to exercise 3.)
8. Prove that the sequence $\{b_n\}$ defined in (18) is Cauchy.
9. Let $\{a_n\}$, $\{b_n\}$, and $\{c_n\}$ be as in (18). Suppose that $\{a'_n\} \approx \{a_n\}$, and let $\{b'_n\}$ be defined as follows:

$$b'_n = \begin{cases} c_n + a'_n - 1, & \text{if } a'_n < 1; \\ c_n, & \text{if } 1 \leq a'_n \leq 2; \\ c_n + a'_n - 2, & \text{if } a'_n > 2. \end{cases}$$

Prove that $\{b'_n\} \approx \{b_n\}$.

10. Complete the proof of Theorem 5.12 by showing that the sequences $\{a_n\}$ and $\{b_n\}$ are Cauchy, $\{a_n\} \approx \{b_n\}$, and if $w = [\{a_n\}] = [\{b_n\}]$ then $f(w) = m$.

11. Complete the proof of Theorem 5.13. (Hint: You may find exercise 7 helpful.)

12. If $x = [\{a_n\}]$ is a real number, then the *absolute value* of x is defined to be the number $|x| = [\{|a_n|\}]$. State and prove the theorem necessary to justify this definition.

13. Prove that it cannot be the case that $|x| + |y| < |x + y|$.

14. Give a Brouwerian counterexample to show that we cannot prove that either $|x + y| < |x| + |y|$ or $|x + y| = |x| + |y|$. (Hint: Find an example for which we cannot determine which of the two possibilities is the case without solving some unsolved problem.)

6

Finitism

The great German mathematician David Hilbert (1862–1943) sought to break the impasse between classical mathematics and intuitionism. To this end, he elaborated a grand project that both developed and knitted together a wide range of philosophical and mathematical ideas.[1]

We can best begin by asking which features of classical mathematics and which of intuitionism Hilbert felt the need to reconcile. Hilbert was unwilling to settle for any solution to the foundations of mathematics that restricted the scope of mathematics. Thus, he did not wish to deprive the working mathematician of reliance on any fundamental principles of inference; in particular, the Law of the Excluded Middle and any inference which it licenses, such as argument by cases, must be preserved. "Taking the principle of excluded middle from the mathematician would be the same," Hilbert said, "as proscribing the telescope to the astronomer or to the boxer the use of his fists."[2] In addition, there should be no doubt about the justification of the classical commitment to the completed infinite: "mathematical analysis, after all, is but a single symphony of the infinite."[3] Hilbert was not prepared to countenance a justification of anything less than the mathematics of the everyday working mathematician.

On the other hand, he himself had questions about the safety and even the intelligibility of infinitary reasoning, and hence about the correctness and even the meaning of the claims of classical mathematics. Intuitionism, while its desire to reform mathematics was unacceptable to Hilbert,

---

[1]   There were of course many important way stations along the path to the full flowering of Hilbert's project as we shall present it. As explained earlier, we are here less interested in describing development than in presenting, in the most interesting light, what developed.

[2]   "The foundations of mathematics" (1927), in van Heijenoort (1967, 476).

[3]   "On the infinite" (1925), in van Heijenoort (1967, 373).

was correct to reflect carefully about the epistemological standpoint of the mathematician. Hilbert believed that the meaning of mathematical statements could not be taken for granted. One must, as intuitionists do and classical mathematicians typically do not, reflect about the nature of our fundamental access to mathematical truth. Only in this way can one come to understand clearly the content of mathematical claims. Furthermore, Hilbert believed that if one does this, then one will indeed find oneself able to justify only a rather restricted collection of forms of inference, certainly ones that are outstripped by those used in classical mathematics and perhaps even by those deemed intuitionistically valid.

The problem thus seems to be the impossible one of justifying the use of classical reasoning on the basis of a far more modest conception of correct inference, that is, of justifying reasoning committed to the completed infinite from a perspective that countenances no such thing. In order to see how Hilbert hoped to eat his infinitary cake and yet have it too, we must turn to what he took to be the fundamental epistemological perspective of the mathematician.

On Hilbert's view, the basic evidence on which mathematics is founded is provided by a faculty of intuition that allows us immediately to take in the symbols of mathematics, which Hilbert describes as "extralogical concrete objects",[4] and to apprehend truths about them and their relationships. "The solid philosophical attitude that I think is required for the grounding of pure mathematics," Hilbert declares, "is this: *In the beginning was the sign.*"[5] Much scholarship has been devoted to fine questions regarding the nature of this intuition (in particular, its connection to sensory experience and to thought), as well as of the objects and truths to which it gives one access. We will not delve too deeply into these matters here. What is clear is that for Hilbert this foundational perspective, which he calls *finitism* and whose meaningful claims are sometimes called *finitary*, or *real*, *propositions*, is not amenable to further justification; in fact, the collection of finitary truths about symbols and finitarily correct principles of reasoning is what constitutes the framework that renders possible "all scientific thought, understanding and communication."[6]

Difficult questions arise concerning the precise scope of finitism. (We shall see later that this imprecision makes elusive any definitive judgment regarding Hilbert's project.) For Hilbert, our fundamental faculty of

---

[4]   "On the infinite," p. 376.

[5]   From Hilbert's "The new grounding of mathematics: first report" (1922), in Ewald (1996, 1121–2).

[6]   Ibid.

intuition, whose operation determines the boundary of finitism, unquestionably supplies us with certain elementary truths about symbols, as well as with principles of reasoning that can be relied on to arrive at yet other truths. These symbols include numerals, by which he understands sequences of strokes, such as |, ||, |||, and so on; these sequences can be abbreviated by the numerals "1", "2", "3", and so on. Certain basic operations can be performed on these numerals, and identities regarding the outcome of such operations can be apprehended. For example, where "+" means the concatenation of numerals, "|| + ||| = |||||" asserts that if one concatenates "||" with "|||" one gets the sequence "|||||"; using the above abbreviations, this assertion might also be rendered in the more familiar manner, "2 + 3 = 5". This is an example of what Hilbert calls a *finitary*, or *contentual*, truth; it is an assertion whose meaning and correctness can be intuitively apprehended through the exercise of the faculty that ultimately provides us with the foundational elements of mathematics. Inequalities will also qualify as finitary: thus "1 + 3 > 5" makes the meaningful, and false, assertion that "|" followed by "|||" is a longer sequence of strokes than is "|||||". No doubt, Hilbert also accepts as contentual all truth-functional combinations of such equations and inequalities: for instance, "2 + 3 = 5 and 5 < 2 × 5". Thus, it is clear that finitism encompasses statements that report numerical identities and computations that do not transcend the finite. It is important to realize that, on Hilbert's view, finitary truths obtain not in virtue of correctly describing some independently existing realm of abstract entities, but rather in so far as they accurately report the results of performing certain basic operations on "concrete objects" that we have access to through the exercise of an epistemologically fundamental faculty of intuition.

But the realm of contentual, or intuitively meaningful, propositions extends beyond statements of the kind treated so far. Most obviously, there are quantificational statements. In the first instance, all bounded quantifications are finitary, for they can be viewed as abbreviations for conjunctions and disjunctions. Thus, "For all $n \leq 100$, $P(n)$" just means "$P(1)$ and $P(2)$ and... $P(100)$"; and "There exists $n \leq 100$, $Q(n)$" is merely short for "$Q(1)$ or $Q(2)$ or... or $Q(100)$."

Of course, unbounded quantification cannot be treated in this manner. In what follows, we shall also consider real those statements of the form $\forall x P(x)$, where $P(x)$ is meaningful from a finitistic point of view. Some care must be taken here, for such universal quantifications will only be deemed contentual if they are properly understood. For instance, if the universal quantifier is understood as ranging over a completed infinite totality of numerals, then the statement is unintelligible to a finitist, since no such totality could be accessed through the faculty of intuition that Hilbert takes to be basic. Instead, the claim must be interpreted as a hypothetical assertion that if any particular numeral n is given, then $P(n)$ will be true for that n. The universal quantification may be viewed as shorthand for $P(1)$, $P(2)$, $P(3)$, and so on, since once one has the hypothetical assertion, one can apply it first to 1, then to 2, and so on. Of course, one must think of this sequence as a potential infinity of propositions.

Three observations are in order here. First, the universal quantifier does not actually appear in Hilbert's finitist mathematics. Rather, generality is there expressed through the use of schemata and special variables.

Second, although we shall here take universal statements to be real, the textual evidence is somewhat mixed. Sometimes Hilbert writes as if such statements are not part of real mathematics, as when he says that a universal "comes to assert something when a numeral is given," suggesting that it does not assert anything until the numeral is given.[7] Sometimes, however, Hilbert writes that universals are *"incapable of being negated"* without giving rise to nonfinitary, or *transfinite*, claims, thereby intimating that they are real statements in their original unnegated form.[8] "We come upon a transfinite proposition," Hilbert says, "when we negate a universal assertion, that is, one that extends to arbitrary numerals," which again suggests that he does not take a universal assertion to be a transfinite proposition.[9] Likewise, Hilbert writes that "a proposition, so long as it is combined with some indication as to its contentual interpretation, is still admissible from our finitist point of view, as, for example, the proposition that always $a + b = b + a$."[10] It seems that Hilbert is here saying that when general statements are interpreted appropriately, they express real propositions. They are, he says, *"finitary propositions* of problematic character."[11]

What makes them problematic, and this leads to our third observation, is that they cannot be meaningfully negated. On the finitist view, an unbounded existential quantification is not so much unwarrantedly asserted on the basis of the denial of a universal claim as it is meaningless. One can speak intelligibly of a specific finite calculation producing an n for which $P(n)$ is true, but in the absence of such a specific calculation it is

[7]   "On the infinite," p. 378.
[8]   Ibid. We will see shortly why Hilbert considered universal statements to be incapable of being negated.
[9]   Ibid.
[10]   Ibid., p. 380.
[11]   Ibid., emphasis added.

not meaningful simply to assert that "some n" makes $P(n)$ true.[12] No interpretation analogous to that given for unbounded universals is available. The only hypothetical claim that one might take an unbounded existential to be making is something like "If one searches through all of the numerals, then one will eventually find an n such that $P(n)$," but of course such a search makes no sense from the perspective of finitism. Furthermore, such an existential claim cannot even be given a partial meaning in terms of what one might expect in the course of such a search. For an unbounded existential claim says nothing about what to expect after any finite amount of checking; it has nothing to tell us about the search until it is completed. This is in sharp contrast to universal claims, where one does get a prediction about what will happen after checking a finite number of cases, namely, that the predicate will apply in all of them. Real propositions, consequently, are not closed with respect to negation: in particular, if one were given a universal statement, then "this communication could not be negated" without arriving at something that lacks finitary meaning.[13]

In general, there are many notions which are at home within classical mathematics but which are not intelligible from the standpoint of finitism. This is correlatively the case for inferences: many classically valid ones outstrip the resources of finitism. For instance, Hilbert is explicit that the familiar inference from $\neg\forall x P(x)$ to $\exists x \neg P(x)$ is not correct (indeed, in light of the above, not always even fully intelligible) from the finitist point of view. Intuitionism, likewise (though it took some time for this fact to become clear to Hilbert and his followers),[14] extends beyond what would be countenanced by Hilbert's foundational stance: for instance, finitism does not have any truck with choice sequences, nor does it take unbounded existential statements to be meaningful.

The most important locating fact, however, is that there is no truth or principle of reasoning acceptable to the finitist that would not be acceptable to both the classical mathematician and the intuitionist; as we shall

---

[12] Hilbert sometimes says that a statement of the form "for some n, $P(n)$" can be used meaningfully as an "incomplete communication" ("*unvollständige Mitteilung*") of a statement specifying how to calculate an n for which $P(n)$ is true. But, again, if one does not have such a calculation in mind, it is not meaningful simply to assert that $P(n)$ is true "for some n." See Hilbert and Bernays (1934, 32).

[13] "The foundations of mathematics," p. 470; note, *à propos* our second observation, that Hilbert treats the general statement as a "communication", that is, presumably, as a contentual assertion.

[14] See, for instance, the Appendix to Bernays (1930) in Mancosu (1998, 263). (See also Mancosu's brief discussion on pp. 167–8.)

see, if this were not the case, finitism could not help resolve the struggle between those who wish to leave mathematics as it is, and those who wish to allay fears about the justification and meaning of mathematical claims.

How can Hilbert hope to effect a resolution? One way of putting his central idea, the way in which Hilbert himself puts it as his project reaches full maturity, is that this reconciliation would be achieved if the consistency of everyday mathematics were established using just the means available within finitism. Hilbert believes that to establish this would be to show that no harm could ever come from the practice of classical mathematics; in particular, fears of paradox would be once and for all put to rest. "And what we have experienced twice," Hilbert declares, "first with the paradoxes of the infinitesimal calculus and then with the paradoxes of set theory, cannot happen a third time and will never happen again."[15] Furthermore, since the consistency proof is to be carried out within finitary mathematics, it would be one that would convince all parties to the dispute. No one, neither the classical mathematician nor the intuitionist, need any longer worry about a contradiction lurking within full-blown infinitary mathematics.

This immediately raises at least two questions: What is "everyday mathematics"? and What is it to establish its consistency? With regard to the first, clearly there is no way to get a handle on mathematics, or any other discipline, one statement at a time. Rather, fundamental principles must be isolated and rules for reasoning must be explicitly outlined; all particular results of mathematics must ultimately be derivable from these principles on the basis of these rules of reasoning. Nothing short of an articulated theory, say an axiomatic theory within which all of classical mathematics can be developed, will permit one even to be able to speak about mathematics as a whole, and to establish that it exhibits certain properties, such as consistency.

Hilbert here turns with gratitude to those, like Frege, who showed how vast areas of mathematics could be systematized on the basis of a small number of assumptions of an arithmetical, set-theoretical, and logical nature. Although these pioneers may have had philosophical goals that differed significantly from Hilbert's, their work provides an axiomatic development of classical analysis that expresses in a surveyable fashion what he calls "the technique of our thinking."[16]

Now if we take some such axiomatic development of mathematics to be given, call it *I*, how is its consistency to be established? The system *I*

---

[15]   "On the infinite," p. 383.
[16]   Ibid., p. 475.

will be consistent just in case there is no contradiction, a statement of the form $P \land \neg P$, that can be derived in it. One way of establishing this is to show that all the axioms of the theory are true; and that all its rules of inference are truth-preserving, that is, applied to true statements, the rules of inference yield only true statements. This would show that every theorem of the system is true; and this of course insures that a contradiction, which is always false, cannot be derived. But clearly Hilbert could not avail himself of this approach, given his goal of neutral reconciliation. For the infinitary claims of classical mathematics can only be understood, let alone be seen to be true, from a perspective that intuitionists are unable to adopt.

Hilbert's innovative solution is to suggest that consistency be established without inquiring into the meaning of the statements in the theory. We can view the axiomatic theory I simply from a formal point of view. That is, we can study the syntactic structure, the form or shape, of the last lines of all possible derivations to ascertain whether any contradiction could appear in the set of theorems. Because any statement can be derived from a contradiction, it suffices to focus attention on just one contradiction; let us choose the statement "$0 = 1 \land 0 \neq 1$", although nothing turns on this particular choice. One can investigate whether this statement is derivable within the axiomatic theory without assuming that one understands what its theorems mean, or that one knows whether they are true. One does this by treating the axiomatic theory as a complex formal object. This involves a syntactic characterization – that is, one that makes reference only to the configuration of symbols – of the system's axioms, of its rules of inference, of what constitutes a derivation, and so on. The question about consistency then takes on a thoroughly syntactical cast: Is there a formal derivation in I whose last line is "$0 = 1 \land 0 \neq 1$"? The problem has just the status of showing that some position on a chessboard is not legal, in other words, that it can never be arrived at from the initial position using the rules of chess.[17] Hilbert thereby replaces a semantic construal of inconsistency (that the theory entails a statement that is necessarily false) by a syntactic one (that the theory formally derives the statement "$0 = 1 \land 0 \neq 1$").

Three observations are again worth making here. The first reports on the historical significance of Hilbert's move. For his approach inaugurated an area of research that continues to flourish: the mathematical study of formalizations of stretches of mathematics. It is sometimes known as *proof theory*, sometimes as *metamathematics*: mathematics

---

[17] Hermann Weyl offers this analogy; see Weyl (1925–7, 137).

turned upon itself to become its own formal object of study. Hilbert urged that mathematicians:

> make the concept of specifically mathematical proof itself into an object of investigation, just as the astronomer considers the movement of his position, the physicist studies the theory of his apparatus, and the philosopher criticizes reason itself.[18]

Regardless of the ultimate success of Hilbert's project, the field of inquiry it spawned continues to yield fruit.

Next, it is important to note that Hilbert's conception of the task that lies at the heart of the foundations of mathematics coheres with his view of the evidence upon which mathematics ultimately rests. For while our faculty of intuition cannot provide us with "immediate experience"[19] of the entities postulated by classical mathematics, it can survey an axiomatic theory, *qua* formal object. Intuition can be presented with finite formal objects such as symbols, sequences of symbols, rules for the transformation of symbols, and so on. Furthermore, the statement of consistency is one that makes finitary sense: for it claims that every formal derivation in the axiomatic system fails to have the sequence of symbols "$0 = 1 \land 0 \neq 1$" as a last line. This is a universal claim, which, as discussed earlier, is to be treated as part of finitary mathematics, although, as already stressed, its meaning there will differ from that assigned to it within classical mathematics. Furthermore, Hilbert believed this finitary claim could be proved using just those means of reasoning acceptable from the finitist standpoint.

Our third observation is that the project of establishing consistency, so conceived, is one that will make sense both to the classical mathematician and to the intuitionist. If the finitist, whose principles of reasoning are accepted by all parties to the dispute, can establish that every derivation fails to terminate with the line "$0 = 1 \land 0 \neq 1$", then this must convince intuitionist and platonist alike that no contradiction is derivable within the formalization of classical mathematics. The project holds out the promise, then, of resolving the dispute between classical mathematicians and intuitionists.

Or does it? There are at least two hesitations one might have about this line of reconciliation. The first is one that can be traced to the French mathematician Henri Poincaré (1854–1912). His objection begins with

---

[18]   The quotation is from Hilbert's "Axiomatic thought" (1918), in Ewald (1996, 1115).

[19]   "The new grounding of mathematics: first report," p. 1121.

the observation that a consistency proof of the kind Hilbert envisaged would need to rely on mathematical induction: after all, the finitary statement that I is consistent makes a general assertion, and induction will probably be needed to justify it. But, Poincaré continues, induction is one of the infinitary principles of inference that is being justified: the classical mathematician uses induction to prove that, say, each element of the completed totality of natural numbers has some particular property. As a consequence, the proposed consistency proof would exhibit a kind of circularity: it would not be convincing to someone who did not already accept the legitimacy of at least some classical inferences.[20]

This objection is immediately avoidable, however, if one distinguishes, as the mature Hilbert did, between induction viewed as a principle employed in metamathematical reasoning and the principle of induction in I. Because the latter is a rule of a formal system, and so makes no claims at all but simply licenses certain manipulations of symbols, there can be no comparison with the meaningful principle used in finitistic reasoning. Furthermore, even if both principles are treated as contentful – that is, as understood by, respectively, the finitist and the classical mathematician – and compared as such, two significant differences emerge. The first is that the predicates to which the classical mathematician will apply induction are of arbitrary quantificational complexity, whereas this is not so for the finitist. Induction applies to all well-defined predicates, and for the platonist these include predicates containing any sequence of universal and existential quantifiers. A finitist, on the other hand, will not countenance predicates involving unbounded existential quantification and will apply induction only to predicates whose sole quantifiers are initial universal quantifiers. It is tempting to conclude from this that the finitist's induction principle is weaker than the classical mathematician's. However, there is a second difference between the two conceptions that makes such a statement misleading. As we saw earlier, the finitist's understanding of the universal quantifier, unlike the classical mathematician's, does not involve commitment to the completed infinite. The finitist's conception is more akin to that of the intuitionist: a statement of the form $\forall x P(x)$ will be justified just in case there is an argument that shows that, for any particular numeral k, $P(k)$ is correct. For this reason, there is no straightforward way to compare the finitist's conception of induction and the classical mathematician's. We saw in Chapter 4 that the classical mathematician can view the validity of induction as that which guarantees that only *finite* iterations of the successor operation will yield a natural number; the realist might also regard induction as a

---

[20]   For instance, see Poincaré (1905), in Ewald (1996).

consequence of an impredicative characterization of the natural numbers. For the intuitionist, and perforce the finitist, induction cannot be understood in either of these ways: the successor operation can only be iterated finitely often – so there is no need for guarantees – and impredicative characterizations are not intelligible. Finitists and intuitionists will instead see induction as that which follows from a correct understanding of the universal quantifier and an appreciation of the constructibility of mathematical objects. As Hilbert says, the two principles that Poincaré assimilates are "essentially different."[21]

There is a second hesitation one might have about Hilbert's projected line of reconciliation between intuitionism and classical mathematics. It will be largely obscured if one thinks that the central dispute concerns the safety of classical forms of inference, in the sense that reliance on them will not generate inconsistency. For it seems clear that the successful prosecution of Hilbert's project will settle that matter: establishing the consistency of classical mathematics using just finitary means should convince all parties concerned that no contradictions will follow from employing infinitary notions and reasoning. Hilbert himself does sometimes cast safety in this sense as the main issue. And he is obviously determined to protect mathematics from the paradoxes.

But there is a different worry that more accurately captures the intuitionists' apprehension about the safety of classical mathematics. For their fundamental concern is that the use of certain forms of inference generates, not contradiction, but unjustified assertions. The general application of the Law of the Excluded Middle, for instance, leads to claims that the intuitionist can find no basis for asserting. The second hesitation, then, is that the successful execution of Hilbert's project would not directly respond to this concern, where a direct response consists in an assignment of fixed, intuitionistically intelligible meanings to individual statements, and a confirmation that the assertions of classical mathematics, when so interpreted, are justified. And of course, it could not: his project is clearly not about assigning any such meanings to the statements of infinitary mathematics, for it proceeds from a perspective that strips classical mathematics of all interpretation. Hilbert's task is not to attach a fixed meaning to the individual statements of classical mathematics on

---

[21] "The new grounding of mathematics: first report," p. 1123. Paul Bernays, in his "On Hilbert's thoughts concerning the grounding of arithmetic" (1922), refers to the principles of the finitist and of the classical mathematician as "narrower" and "wider," respectively, but it is clear from his discussion that they operate so differently that any direct comparison of strength is spurious (Bernays, 1922; in Mancosu, 1998, 221). See also the penultimate paragraph of Bernays (1923).

the basis of which all parties can see that they are correctly asserted. If the intuitionist's worry about safety is responded to by Hilbert, it must be in a different way.

To see how Hilbert seeks to answer the intuitionist, we must first consider some natural questions about the relationship between the set of formal theorems of *I* and the collection of finitary truths. If *Q* is a true real statement that is quantifier-free, then its truth can be established simply by means of a calculation, and it is plausible to assume that we can carry out a derivation of *Q* based on this calculation in a formalization of infinitary mathematics. Thus, using the shorthand *S* ⊢ *P* to mean that *P* is formally derivable from the axioms of the formal system *S* using just the rules of inference of *S*, we shall assume

(1) If *Q* is a quantifier-free real statement, then if *Q* is finitarily justifiable then *I* ⊢ *Q*.[22]

We will consider the case where *Q* is not quantifier-free later; (1) is all that is needed for the present discussion. There are of course many statements which are derivable in *I* which are not correct from the point of view of finitism, since there are many formal theorems that are not even intelligible to the finitist, for instance unbounded existential assertions: in this sense, *I extends* finitary mathematics. What is not clear is whether in *I* one can formally derive any *finitary* statements that are not finitarily correct. That is, might the infinitary resources of classical mathematics extend, beyond the bounds of the finitarily correct, the collection of finitarily meaningful statements that can be established? We can restate this question by asking whether *I* is a *conservative extension* of finitary mathematics with respect to the collection of real statements; that is, whether (2) holds:

(2) If *Q* is a real statement, then if *I* ⊢ *Q* then *Q* is finitarily justifiable.

An affirmative answer in effect would mean that any real, or contentual, statement that could be derived in classical mathematics could already have been justified using only the resources of finitary mathematics.

Let us consider an example. Fermat's Last Theorem states that for all positive integers *x*, *y*, *z*, and *n*, if $n > 2$ then $x^n + y^n \neq z^n$. This is the famous theorem – for over three hundred years, conjectured to be true yet remaining unproved – that the British mathematician Andrew Wiles

---

[22] Or, more accurately, *I* proves the formal statement corresponding to *Q*. In what follows, we shall simply use *Q*, and so on, with systematic ambiguity.

(1953–) finally established in 1994. Notice that the theorem has the form of a universal statement about finite calculations, and is therefore finitarily meaningful. But Wiles' proof relies on a great deal of sophisticated infinitary mathematics that is not finitarily meaningful. How can we be sure that the conclusion of Wiles' proof is correct? Of course, we can verify any particular instance of the theorem by plugging in numerals for x, y, z, and n and performing the relevant computations; for example, we could confirm by means of a finite calculation the truth of the statement:

(3) $5,874^{167} + 104,134^{167} \neq 380,434^{167}$.

Going through this laborious calculation is one way of checking this particular consequence of Wiles' infinitary reasoning. But another way would be to establish that infinitary mathematics is a conservative extension of finitary mathematics: this fact would provide a general guarantee that all calculations like this would turn out as predicted by Wiles' proof.

If infinitary mathematics could be shown to be such a conservative extension, then there would be a clear sense in which its safety would have been demonstrated to those who, with the intuitionists, worry about the fundamental evidence for infinitary claims, indeed about their basic intelligibility: the fact that any particular use of infinitary mathematics to generate finitary truths could in principle be eliminated in favor of finitary reasoning shows that it can be trusted to give only correct finitary results. This, then, is how Hilbert seeks to respond to the intuitionists' worry. Hilbert does not try to defend classical mathematics by showing that the intuitionists are wrong about the difficulties in justifying or understanding particular claims about the actual infinite. Quite by contrast, he seeks to justify infinitary statements instrumentally and holistically. The justification is instrumental in that infinitary statements are to earn their keep through their utility in establishing correct real statements, and it is holistic in that this usefulness is to be established by examining infinitary statements not one at a time but *en masse*, as a complex integrated deductive structure. Hilbert's goal would be achieved if he could show that any particular use of infinitary mathematics to derive a real statement could in principle be eliminated in favor of a completely finitary justification of that statement. Infinitary derivations of real statements would remain useful, however, because they are often much shorter, perhaps even to the point of permitting the derivation of results that could not in practice have been arrived at by more direct means; and also because they help to systematize our knowledge of real statements by revealing connections that would otherwise be hidden by the more laborious, if direct, finitary justifications. Hilbert explicitly draws the analogy to one way of looking at theories in natural science:

> What the physicist demands precisely of a theory is that particular propositions be derived from laws of nature or hypotheses solely by inferences, hence on the basis of a pure formula game, without extraneous considerations being adduced. Only certain combinations and consequences of the physical laws can be checked by experiment – just as in my proof theory only the real propositions are directly capable of verification. The value of pure existence proofs consists precisely in that the individual construction is eliminated by them and that many different constructions are subsumed under one fundamental idea, so that only what is essential to the proof stands out clearly; brevity and economy of thought are the *raison d'être* of existence proofs.[23]

Hilbert's hope, in short, is to justify classical mathematics to intuitionists by demonstrating its dispensable utility when dealing with finitary statements. As one commentator put it, infinitary mathematics could then be viewed as a collection of "gadgets to make life smoother."[24] But how does this relate to Hilbert's project of searching for a consistency proof?

We are finally in a position to see how the second hesitation about the significance of Hilbert's proposed consistency proof can be dispelled: for it is a fundamental fact that if we can give a finitary proof that I is consistent, then I is a conservative extension of finitary mathematics. To see why this is so, suppose that Q is a quantifier-free real statement such that $I \vdash Q$; we want to show that Q is finitarily justifiable. Now, since Q is quantifier-free, it can either be confirmed or refuted by a finite calculation. Assume that Q is finitarily refuted. Given (1), it follows then that $I \vdash \neg Q$; hence, I is inconsistent. Thus, we are finitarily justified in asserting that if I is consistent, then Q is true. Combining this with the finitary proof of I's consistency yields a finitary justification of Q.

Let us assume now that Q is a real statement that expresses generality, say a statement of the form $\forall x P(x)$, where $P(x)$ is quantifier-free; we shall show that if it is finitarily provable that I is consistent and if $I \vdash \forall x P(x)$, then $\forall x P(x)$ is finitarily justifiable.[25] Let k be any numeral. Since

---

[23]   "The foundations of mathematics," p. 475. See also Weyl (1925), in Mancosu (1998), p. 140.

[24]   Kreisel (1983, 209).

[25]   Once again, the universal quantifier here cannot be understood to range over a completed infinite totality. The finitary justification that follows in the text reflects this. The French mathematician Jacques Herbrand (1908–31), describing the principles of reasoning acceptable to an adherent of Hilbert's program, writes

$I \vdash \forall x P(x)$, $I \vdash P(k)$ (by universal instantiation). But $P(k)$ is a quantifier-free statement, so we can use the reasoning in the previous paragraph to give a finitary justification of $P(k)$. Since k was arbitrary, this constitutes a finitary proof of $\forall x P(x)$.

What the last two paragraphs establish is that (2), in its full generality, follows from the assumption that we have a finitary proof of the consistency of I. Furthermore, the reasoning from consistency to conservation is fully acceptable to an intuitionist. Therefore, an intuitionist must accept that a finitary consistency proof of I would suffice to establish the reliability of classical mathematics for all finitarily meaningful statements. Given such a proof, the intuitionist would no longer have any reason to eschew the use of classical mathematics for the derivation of finitary truths: for any such derivation could be transformed into a finitary justification of the relevant truth. The method has essentially already been given, but it might be worthwhile to pursue an example. Let us return to the classical derivation of the finitary statement (3) which proceeds *via* Wiles' infinitary proof of Fermat's Last Theorem, and let us assume that we are in possession of a finitary proof of the consistency of I. Here is how we can employ this derivation and this proof to provide a finitary justification of (3). Since (3) is quantifier-free, it can be either confirmed or refuted by a finite calculation, so we may assert that it is either true or false. Assume that it is false. Since the negation of (3) is finitarily justifiable (by carrying out the relevant computation), it is, by (1), derivable in I. But (3) is also derivable in I; hence a contradiction is as well. But there is, we are here assuming, a finitary proof of the consistency of I, which rules out such a derivation. Thus, we can infer that our original assumption that (3) is false is incorrect. This is a finitary justification of the truth of (3). Of course, there is nothing special about the particular numerals that appear in (3). Similar reasoning could be applied to any instance of Fermat's Last Theorem, and this observation constitutes a finitary proof of the theorem.

A finitary proof of consistency therefore holds out the extraordinary prospect of vindicating both the classical mathematician and the intuitionist, and in so doing reconciling each to the other. It would legitimate classical forms of reasoning, by showing that they could not be used to

the following about the finitary justification of universal claims: "we never consider the totality of all the objects x of an infinite collection; and when we say that an argument (or a theorem) is true for all these x, we mean that, for each x taken by itself, it is possible to repeat the general argument in question, which should be considered to be merely the prototype of these particular arguments" (see van Heijenoort, 1967, 622, n. 3).

derive any real but finitarily unjustifiable statement, without thereby contradicting the intuitionist's contention that classical infinitary statements do not possess general meanings.

This was the project to which Gödel set out to contribute. In his attempt to advance Hilbert's program, Gödel soon proved two fundamental results in mathematical logic that instead have created considerable doubt about whether Hilbert's goal can be achieved. These are known as Gödel's Incompleteness Theorems, undoubtedly the most celebrated and fecund results in logic. In the next chapter, we shall state these theorems carefully and prove them in considerable detail. But for our purposes here, the following formulations will be adequate. Let us first define a formal system S to be *syntactically complete* just in case, for every statement Q, either $S \vdash Q$ or $S \vdash \neg Q$. The first of the two central results in Gödel's 1931 paper[26] can be loosely stated this way:

First Incompleteness Theorem. If S is a sufficiently powerful formal system, then if S is consistent then S is syntactically incomplete.

The intuitive idea behind Gödel's proof is that if S is a sufficiently powerful formal system, then one can derive in it statements that can be viewed as saying something about the syntactical properties of S itself. The possibility of doing this depends on two facts. First, it is possible to represent syntactical properties of a formal system in numerical terms: syntactical objects can be represented by numbers, and syntactical properties can be represented as number-theoretic properties. Second, basic claims about the natural numbers can be established in a sufficiently strong formal system. A consequence of these two facts is that a sufficiently powerful formal system can derive statements that "say" something about the system itself. Gödel considered a formal statement that could be viewed as asserting its own unprovability within some sufficiently rich formal system S. Call this statement, G_S, the Gödel sentence of S; intuitively, G_S says "I am unprovable in S." Is G_S derivable in S? Assume that S is a *sound* formal system; that is, if a sentence is derivable in it, then it is true in the intended interpretation. We see then that G_S is not derivable in S. If it were then it would be false, since it says of itself that it is not derivable in S. But then S would not be sound, contradicting our assumption. Furthermore, since G_S is unprovable in S and says that it is, it is true. Consequently, the negation of G_S is false, and so, by the soundness of S, it is not derivable in S either. It follows that G_S is neither provable nor refutable in S. It is thus an *undecidable*

---

[26] Gödel (1931).

statement in *S*, and its existence demonstrates that *S* is not syntactically complete.

In the next chapter, we will see, following Gödel's original proof, how the syntactic incompleteness of *S* can be demonstrated without having to make the semantic assumption of soundness: it will be replaced by the purely syntactical property of consistency, and the argument to incompleteness will make no reference to truth or any other semantic notion. Here, however, we will focus on the result itself and its consequences for Hilbert's project.

Incompleteness was certainly not expected by Hilbert or by others working to establish the consistency of classical mathematics. The conviction about completeness seems to have been a consequence of three distinct beliefs: first, the correctness of the Law of the Excluded Middle; second, the view that every mathematical truth is provable;[27] and third, the belief that one can capture within one formal system all the basic conceptions and forms of reasoning of infinitary mathematics. It follows from these three that there is a formalization of classical mathematics within which every well-formed statement in the formal language is either derivable or refutable. And this is just what Gödel's First Incompleteness Theorem rules out: no consistent formal system that is sufficiently powerful (as it would have to be if classical mathematics could be represented in it) is complete. Clearly, Hilbert was in error about something.

The First Incompleteness Theorem does not, however, indicate which of these three beliefs must be abandoned, but only that at least one of them must be. And while Hilbert may have held all three, they are not all part of what we have been describing as Hilbert's project. Arguably, the first and third beliefs are integral to the overall framework of Hilbert's attempt to reconcile the narrow evidential perspective of finitism with the broad scope of classical mathematics. Infinitary mathematics crucially depends on the Law of the Excluded Middle, so its abandonment would not leave much to be reconciled. Also, without the assumption that one can capture the principles of classical mathematics within one formal system, Hilbert's consistency proof, were it attainable, would not provide the hoped for reconciliation: questions about the safety of the nonformalized conceptions or principles of reasoning would remain. By

---

[27]   Hilbert held that every mathematical truth is in principle knowable: famously, he declared that "in mathematics there is no *ignorabimus*." From his "On the infinite," p. 384. Hilbert even had had inscribed on his tombstone "*Wir müssen wissen. / Wir werden wissen.*": We must know. / We will know. (See Reid, 1970, 220.)

contrast, it seems that Hilbert's belief that all true mathematical propositions are knowable can be abandoned without jettisoning either the problem he sought to address or his proposed solution to it. Some mathematical truths might not be provable, and yet it might still be the case that there are no real propositions that can be justified on the basis of infinitary, but not finitary, reasoning. In sum, the phenomenon of incompleteness is not in itself a threat to Hilbert's project as here outlined, although it might be to certain other positions Hilbert independently maintained: the point of Hilbert's project is not that infinitary formal mathematics can solve all of our mathematical problems, although Hilbert may well have believed that but, rather, that if it solves a problem, then that solution can be trusted to be correct.

Incompleteness does, however, point to a certain limitation of Hilbert's project when viewed as an attempt to reconcile classical mathematics with intuitionism. One might have already suspected as much on account of the restricted nature of the conservation result described above. We saw that establishing, by finitary means, the consistency of I would suffice to prove (2), that I is a conservative extension of finitary mathematics with respect to finitary statements. There are, however, many statements intelligible to an intuitionist that are not finitary (although, as noted earlier, this was not something immediately understood by Hilbert and his followers). For instance, we have seen that all existential statements fail to be finitary; likewise, any universal quantification $\forall x P(x)$, where $P(x)$ is not decidable. Can infinitary reasoning be trusted not to lead us astray with regard to *these* statements? We cannot look for reassurance to the informal proof of (2) given above, for it does not carry over to these nonfinitary statements. Consequently, we lack an argument that a finitary consistency proof of I shows I to be a conservative extension of intuitionistic mathematics with respect to the intuitionistically meaningful statements. This is, of course, not the same as having an argument that consistency does not entail this conservative extension result.

Gödel's discovery of incompleteness supplies us with just such an argument. Assume that we have a finitarily acceptable proof of the consistency of I. The incompleteness result now guarantees that G_I is not provable in I. In fact, Gödel's reasoning shows explicitly how a derivation of G_I in I could be transformed, by means of a finite calculation, into a proof of a contradiction in I. It follows that we have a finitary proof of the nonprovability in I of G_I; that is, we have a finitarily, and hence also intuitionistically, acceptable proof of G_I's truth, since what G_I claims is that it is unprovable in I. Consequently, $\neg G_I$ is intuitionistically incorrect. Let us consider now the formal system I' which is the same formal system as I, except with one additional axiom, $\neg G_I$. The consistency of I' is likewise finitarily justifiable. For if I' were inconsistent, then in I we would be able to prove that $\neg G_I$ implies a contradiction, which would then constitute a proof (by *reductio ad absurdum*) of G_I in I. But we have a finitary proof that G_I is not derivable in I, so this is impossible. Thus, we have finitarily established that I' is consistent. And yet I' exhibits a certain peculiarity, for it proves the statement, $\neg G_I$, which we just saw is, under present assumptions, intuitionistically refutable! The phenomenon of incompleteness thus opens the possibility that a formal system is proved consistent by finitary means even though one can derive within it a statement that is intuitionistically false. The French philosopher and mathematician Blaise Pascal (1623–62) was thus at least half right when he declared that "Contradiction is not a sign of falsity, nor the lack of contradiction a sign of truth."[28]

This last observation does not cast into doubt our earlier argument that the truth of (2) follows from the existence of a finitary consistency proof. In order to see why, we must consider a little more carefully the nature of G_I, the Gödel sentence of I. This statement, recall, can be viewed as claiming that it itself is unprovable in I. To claim that a statement is unprovable in some formal system is to claim that no derivation in the system has that statement as a final line; put otherwise, it is to claim that each derivation in the system fails to have that statement as a final line. Consequently, G_I in effect says that all derivations in I fail to terminate with G_I. In the following chapter, we shall examine more precisely how G_I is constructed, but for now the important point is that G_I is a universal statement of the form $\forall x \neg P(x)$, where $P(x)$ is a very complicated open sentence of the formal language that "says" that x is a derivation in I whose last line is G_I. As a consequence, the Gödel sentence's negation is not finitarily intelligible: $\neg G_I$, although intuitionistically false, is finitarily meaningless, and hence its derivability in a system that is assumed to have been proven consistent by finitary means does not conflict with our claim that (2) follows from such a consistency proof.

There is no denying, however, that this particular consequence of incompleteness does point to a certain limitation of Hilbert's project, understood as an attempt to reconcile classical mathematics with intuitionism. For it reveals that a finitary consistency proof should not convince an intuitionist to trust infinitary arguments to *all* intuitionistically meaningful conclusions. In particular, derivation of an unbounded existential statement in a formal system proved consistent by finitary

---

[28]   Auden and Kronenberger (1966).

means does not entail that statement's intuitionistic correctness. At most, and this is not without interest, a finitary consistency proof would guarantee that any justification within classical mathematics for a proposition that is meaningful from a finitist perspective is merely an unnecessary, although perhaps simplifying, detour through the infinite.

But even this was not to be. For shortly after Gödel proved his First Incompleteness Theorem, he realized that it has a remarkable consequence that makes it unlikely that such a consistency proof could ever be found. In his seminal paper, Gödel established that if a sufficiently powerful formal system S is consistent, then it fails to prove its Gödel sentence. But the Gödel sentence of S says that it is unprovable in S. Hence, the consistency of S entails that G_S is true. What Gödel noted is that this line of reasoning is itself formalizable within S; that is, within S one can derive a conditional statement whose antecedent expresses that S is consistent and whose consequent is simply the statement G_S. It follows that if S can prove its own consistency, then, by *modus ponens*, within S one can derive G_S. But the First Incompleteness Theorem tells us that if S can derive G_S, then S is inconsistent. Contrapositively, if S is consistent, then S cannot prove its own consistency:

Second Incompleteness Theorem. *If S is a sufficiently powerful formal system, then if S is consistent then S cannot prove its own consistency.*

In the following chapter, we will be more explicit about the content of this theorem and its proof. But here we are interested in the consequences it has for Hilbert's project.

These are substantial. In the first instance, this theorem makes it very unlikely that a finitary proof of the consistency of I will ever be found. Obviously, this was very bad news indeed for Hilbert, whose entire project consisted in the search for such a proof. This assessment is based on the following assumption:

(1*)   If Q is a real statement, then if Q is finitarily justifiable then $I \vdash Q$.

This differs from (1) above in that the latter makes a claim only about quantifier-free real statements. Hilbert's reasoning to the conclusion (2) above requires only assumption (1). But the additional claim that (1*) makes, that finitary reasoning to a general conclusion is likewise captured in a formalization of classical mathematics, is overwhelmingly plausible. Indeed, if it failed to hold it would be doubtful whether finitary mathematics could provide a foundational standpoint accepted by both classical and intuitionistic mathematicians. If one does grant this additional claim, however, then a finitary proof of the consistency of I will be echoed by a formal derivation in I of a statement that asserts I's consistency. But the Second Incompleteness Theorem reveals that this cannot occur so long as I is consistent. The theorem therefore guarantees that if I is consistent, then one will have to choose: either not all finitary reasoning is formalizable within I, or there is no finitary proof of the consistency of I. The first is quite unlikely, while the second spells the unrealizability of Hilbert's project.

The matter is not amenable to a definitive resolution because, as already noted, Hilbert was not completely explicit regarding the scope and resources of finitary reasoning. Still, it is a challenge to explain why any formal system that systematizes a sufficiently powerful stretch of classical mathematics must fail to capture some reasoning that is justified from the finitist standpoint. And also, as just noted, should some finitary reasoning not be formalizable within I, it would be doubtful whether Hilbert's project could do the kind of foundationally reconciling work one might have hoped for. Consequently, Gödel's Second Incompleteness Theorem is usually viewed, and rightly, as casting a pall over Hilbert's project that is not likely to be lifted.

It may be instructive to consider the application of Gödel's Incompleteness Theorems to the formal system PA. We remarked a moment ago that it is doubtful that there might be some finitary reasoning that could not be expressed in I, and the same might be said of PA. However, something *is* failing to be captured by PA: in particular, we can reason to the conclusion that G_{PA}, the Gödel sentence of PA, is true, but something about this reasoning cannot be expressed within PA itself. It seems strange that this should be so, that we can reason to the truth of G_{PA} – a universal statement interpretable as a claim about the natural numbers – whereas PA, whose business it is to formalize our knowledge of the natural numbers, cannot. Our reasoning is simply that G_{PA}, in the light of the encoding of syntactic claims into number theory, can be viewed as expressing the claim that the statement G_{PA} is itself unprovable in PA; and this claim must be true, otherwise one would be able to prove in PA a false statement. This reasoning to the truth of G_{PA} thus crucially depends on the assumption that PA is a sound formal system, that is, one in which derivability entails truth in the intended interpretation. If we can conclude that G_{PA} is true but PA cannot prove it, that must be because we can somehow apprehend that PA is sound while PA itself is incapable of establishing this.[29]

---

[29]   If mechanists are right, and the mind can in some sense be modeled by, say, the formal system PA, then, assuming we are consistent, we *cannot* conclude that

And how *do* we recognize that PA is sound? To some classical mathematicians, a natural answer is that we are able to apprehend mentally the intended model, $\mathfrak{R}$, for the language of PA. $\mathfrak{R}$'s domain, or universe of discourse, is the completed totality of natural numbers, and it provides an interpretation of the language of PA under which all of PA's axioms make true claims about the elements of this domain, and all of PA's rules of inference are truth-preserving. It follows by induction that every theorem of PA makes a true claim about the collection of natural numbers.[30] We will have more to say about what a model is in the following chapter. But for now, the important point is that, for such a classical mathematician, our conviction that G_{PA} is true is ultimately based on our apprehension of the intended model of PA.

On this way of looking at the matter, the unprovability in PA of G_{PA} signals that our grasp of the intended model is not fully articulated by the formal system PA: the resources of PA are not sufficient to prove the system's Gödel sentence, even though we, on the basis of our grasp of the intended model, are capable of seeing that it is true. Our concept of arithmetic truth cannot be identified with the notion of provability in PA.

But the limitation appears not to be specific to PA. Consider what happens if we attempt to remedy the situation by augmenting PA by adding its heretofore unprovable Gödel sentence as an axiom; that is, consider the formal system $PA^+ = PA + G_{PA}$. PA^+ is obviously capable of proving the Gödel sentence of PA, so one might think that it captures everything there is to capture about our understanding of the natural numbers. However, PA^+, being a sufficiently powerful formal system, is likewise subject to the First Incompleteness Theorem: assuming that it is consistent, it cannot prove *its* Gödel sentence, the sentence that asserts that it is unprovable in PA^+, even though we can recognize that it too is true. Clearly, although we can continue this process of augmentation *ad infinitum*, Gödel's theorem shows that it is to no avail: each resulting formal system will be incomplete in failing to prove some statement that

G_{PA} is true. We could still conclude that *if* PA is consistent, *then* G_{PA} is true, for this conditional claim lies within the purview of PA itself. The consistency of PA seems so much to go without saying that we are barely conscious of the need for this assumption in establishing the truth of G_{PA}. For some further discussion, see George and Velleman (2000). For the sake of further discussion, however, we shall henceforth assume that we *can* apprehend the truth of G_{PA}. Our question now is what to make of that fact.

[30] The reader might continue to wonder why the above informal inductive argument for the soundness of PA cannot be formalized in PA. Once we reach Theorem 7.20 in the following chapter, it will be clear precisely why this argument cannot be expressed in PA.

we can recognize to be true (namely, that system's own Gödel sentence) by virtue of our grasp of the intended model, $\mathfrak{N}$. What this easily suggests is that our grasp of the intended model is something that forever eludes explicit articulation. Try as we might to describe in a formal system the model that we all apparently succeed in grasping, we cannot: it seems to outrun all attempts to describe it. We have, it appears, a mental conception that is incapable of explicit linguistic description.

That this is possible was something that Brouwer had insisted on: mathematical activity is mental and, he also claimed, language is necessarily inadequate to communicate fully such mental goings-on. But there is still something here that would trouble Brouwer, and indeed any intuitionist. Idealism claims that there is no gap between reality and our justified beliefs about it. For intuitionists, the idealists of mathematics, there is no intelligible distinction between what is mathematically true and what one can prove to be true, no justification-independent world of mathematical facts relative to which our beliefs are true or false. For intuitionists, truth is not independent of proof, and yet this independence is precisely what the above situation might suggest. No system of mathematical proof is able to capture all arithmetical truths, let alone all mathematical truths. Consequently, truth must transcend proof; there must be a mathematical reality, in this case the totality of natural numbers that we apprehend when we grasp the intended model, that systems of proof can describe with ever greater accuracy but can never completely characterize. It seems not only that Gödel's work brings down the most promising attempt to reconcile classical and intuitionistic mathematics, but also that it actually weighs in on the side of a realist construal of mathematics.

So matters naturally appear from a realist's perspective. That such a view of Gödel's theorems is not forced upon one, however, can be made clear by considering what might initially seem a surprising fact: that intuitionists will likewise assert the consistency of PA, and so too the truth of PA's Gödel sentence. How can this be? An intuitionistic proof of the consistency of PA will have two parts: the first establishes that if a contradiction is provable in PA, then a contradiction is provable in HA, a formalization of intuitionistic arithmetic; and the second shows that HA is consistent. We have already seen how the first half of this proof would proceed: for we saw in the previous chapter that, as a result of Gödel's work in 1933, any sentence which contains neither disjunctions nor existential quantifiers and which is provable in PA is also provable in HA. The intuitionist will conclude that if HA is consistent, then PA is.

This alone does not provide the intuitionist with a guarantee that PA is consistent. For that conclusion, the intuitionist must be convinced that

HA, a particular formal system, will never prove a contradiction. The intuitionist might offer the following informal proof to that effect. Each axiom of HA is, from the intuitionist's point of view, true: that is to say, each axiom makes a claim that is justifiable on the basis of the intuitionistic construction of the natural numbers. Likewise, each rule of inference of HA preserves intuitionistic truth, or correct assertability. By induction, then, it follows that every theorem of HA is a justifiable assertion about the natural numbers, that is to say, true. Clearly, this proof bears a schematic resemblance to the classical mathematician's proof of the consistency of PA. But where the classical mathematician appeals to a notion of truth in a model whose universe of discourse is a completed infinite totality, the intuitionist's conception of a true statement is akin rather to the notion of an assertion justifiable by reference to a conception of the natural numbers as a potentially infinite collection.

Putting these two halves together, we arrive at a proof of the consistency of PA from an intuitionistic standpoint, a perspective from which there is no distinction to be drawn between truth and provability. Consequently, that one can recognize PA to be consistent cannot by itself signal a vindication of classical over intuitionistic mathematics: each perspective can, in its own way, come to that recognition.

If PA is consistent, we know by the Second Incompleteness Theorem that it cannot prove its own consistency. Consequently, assuming PA to be consistent, something about the intuitionist's reasoning must not be representable within PA. To the intuitionist, the natural conclusion to draw is that the intuitionistic notion of proof is not captured by PA: for the intuitionist, it *is* provable that PA is consistent, whereas this is something that PA cannot prove. The concept of being intuitionistically provable transcends that of being provable in PA.

Of course, as before, the resources of PA can be expanded to include the intuitionistically provable but previously formally underivable assertion that PA is consistent. But this expansion will give rise to a new formal system in which some other intuitionistically provable assertion will not be derivable. Furthermore, the Incompleteness Theorems tell us that this process of extending and strengthening formal systems is a never-ending one. What all this shows, intuitionists would hold, is that the informal notion of proof with which they operate and in terms of which they understand mathematical statements transcends the notion of derivability in any particular formal system: no sooner do we think we have a formal system whose notion of provability encompasses that of intuitionism than we can articulate an intuitionistically correct proof that cannot be captured in that formal system.

But is this not a version of the very lesson that we just suggested classical mathematicians draw from Gödel's work? Did we not say earlier that for them truth transcends provability in any particular formal system? Yes, but the crucial difference arises in how intuitionists and classical mathematicians go on from this common conclusion. By way of comparison, recall that both classical mathematician and intuitionist likewise agree that no rule for enumerating real numbers will enumerate all of them. The difference between them is that the intuitionist is willing to assert only that, given any rule for enumerating real numbers, one can construct a real number that would not be so enumerated; while the classical mathematician believes that this infinite process of extension is in some sense capable of being completed to yield an actual set that contains all the real numbers. Analogously, the intuitionist asserts only that given any sound formal system for proving propositions, one can construct a proposition which, though provable, is not a theorem of that system; while the classical mathematician treats this process of extension as capable of being completed, with the result a determinate collection that contains every single justifiable claim. For the intuitionist, the concept of correct, or justifiable, claim, is, like the concepts of set, of natural number, or of real number, indefinitely extensible: Gödel's theorems show that any formal characterization of the notion of mathematical justification will provide the basis for the description of a justified claim not previously recognized to be such. For the classical mathematician, by contrast, there is a determinate collection which results "after" the never-ending process of extension is completed. The insistence that there is a completed collection of correct claims that is distinct from the collection of claims provable in any given formal system is no doubt of a piece with the classical mathematician's view that there is a distinction between truth and provability. An intuitionist will insist that there is no such thing as "the limit" of the process of extending descriptions of what it is to be a correct mathematical claim: being an infinite process, it cannot be completed. All there are are actual characterizations of our concept of correctness, of being legitimately proved, any given one of which can be extended to a more encompassing one. From the intuitionist's perspective, the classical mathematician's conception of truth is bound up with an unintelligible extrapolation from the never-ending process of extending our conception of a justified claim.[31]

It seems, in sum, that, while Gödel's Incompleteness Theorems make it very unlikely that anything like Hilbert's grand project of reconciliation can be accomplished, they do not favor either of the fundamental philo-

---

[31] For an important and related discussion, see Dummett (1963).

sophical perspectives on mathematics that Hilbert sought to harmonize. The two perspectives, still unreconciled, can understand and accommodate Gödel's results, each in its own distinctive and self-reinforcing manner.

Exercises

1. In our informal discussion of Gödel's Incompleteness Theorems in this chapter, we considered a sound formal system S, and a sentence G_S that could be viewed as saying "I am unprovable in S." Let us consider instead a sentence R_S that says "I am refutable in S," where a sentence P is said to be *refutable* if $\neg P$ is provable.[32]
    (a) Is $\neg R_S$ provable in S? (In other words, is R_S refutable in S?)
    (b) Is R_S true or false?
    (c) Is R_S provable in S?
2. A certain island is inhabited by knights, who always tell the truth, and knaves, who always lie. Furthermore, some of the knights have proven themselves to be knights, and are known as *established knights*, and similarly some of the knaves are *established knaves*.[33] (You might want to compare the three exercises below to the sentences G_S and R_S discussed above, and exercise 15 of Chapter 7.)
    (a) An inhabitant of the island says "I am not an established knight." Can you tell whether he is a knight or a knave, and whether or not he is established?
    (b) An inhabitant of the island says "I am an established knave." Can you tell whether he is a knight or a knave, and whether or not he is established?
    (c) An inhabitant of the island says "I am an established knight." Can you tell whether he is a knight or a knave, and whether or not he is established?
3. Let us say that an English phrase that picks out a particular natural number is a *name* for that natural number. For example, the phrase "the only even prime number" is a name for the number two.[34]

---

[32] This problem is based on an idea from Smullyan (1982, 187–8).
[33] This problem is based on an idea from Smullyan (1978, 225–6).
[34] This exercise is based on a paradox that was communicated to Bertrand Russell by G. G. Berry, of the Bodleian Library at Oxford University. You might want to compare this exercise to exercise 23 in Chapter 7.

(a) Show that there are only finitely many numbers that have names that are 12 words long, and use this to establish that there is a smallest number that does not have such a name.

(b) By part (a), the phrase "the smallest natural number that does not have a 12-word name" is a name for a natural number. What is wrong with this?

(c) As part (b) shows, some phrases that appear to be names for natural numbers lead to paradoxes. Suppose that we have a procedure for verifying that a given phrase is a nonparadoxical name for a natural number. We will say that names that have been verified by this procedure are "established names." Is the phrase "the smallest natural number that does not have a 13-word established name" a name? Is it established?

7

The Incompleteness Theorems

The Incompleteness Theorems apply to a wide variety of formal systems, but to simplify the discussion we will focus for most of this chapter on one formal system. The formal system we will use is a version of the Peano Postulates, and is often referred to as *Peano Arithmetic*, or *PA* for short. We have already seen, in Chapters 2 and 3, that the Peano Postulates play a central role in the foundations of mathematics, and we will see in this chapter that certain properties of PA play central roles in the proofs of the Incompleteness Theorems. The language of PA includes, in addition to the usual symbols of logic ("∧", "∨", "¬", "→", "↔", "∀", "∃", "(", ")", and "="), the symbols "0" (zero), "S" (for the successor operation), "<", "+", and "·", and variables "x_1", "x_2", "x_3", ... (which will range over the natural numbers). The axioms are the usual axioms of first-order logic, together with sentences in this language that express the Peano Postulates and sentences saying that the symbols "<", "+", and "·" satisfy the usual definitions of the less-than relation and the operations of addition and multiplication.

The details of how formal systems operate will be more important in this chapter than they have been in previous chapters, so it may be worthwhile to explain the workings of PA in some detail. We begin by defining the two kinds of well-formed expressions in the language of PA, *terms* and *formulas*.

DEFINITION 7.1.   The *terms* of the language of PA are the expressions generated by the following rules:
  (i)   "0" is a term.
  (ii)   Every variable is a term.
  (iii)   If t and u are terms, then so are the expressions St, $(t + u)$, and $(t \cdot u)$.[1]
  (iv)   Nothing else is a term.

---

[1]   There is a slight abuse of notation here. The letters "t" and "u" are not themselves symbols in the language of PA, but are rather letters that we are

For example, by rule (i) of Definition 7.1, "0" is a term. Therefore, by rule (iii), so is "$S0$". By rule (ii), "x_3" and "x_{12}" are terms, so by rule (iii), so are "$(x_3 + S0)$" and "$(x_{12} \cdot (x_3 + S0))$". Note that according to Definition 7.1, the parentheses are necessary in these examples. For example, "$x_{12} \cdot x_3 + S0$" is not a term. Intuitively, terms are expressions that denote natural numbers when values are assigned to the variables, and the rules about parentheses guarantee that the meanings of terms are unambiguous. However, it is important to realize that Definition 7.1 makes no reference to the meanings of the symbols; whether or not an expression is a term is determined entirely by how the symbols are arranged, and not what the symbols mean. In fact, this will be true of many of our definitions in this chapter, and this is an important part of what we mean when we say that PA is a formal system. Although it is often helpful to think of expressions in the language of PA as being meaningful, all concepts about PA that we will use in the proofs of the Incompleteness Theorems will be defined in terms of syntactic properties of arrangements of symbols, with no reference to any meanings that might be assigned to these symbols.

The terms "0", "$S0$", "$SS0$", "$SSS0$", ... will be especially important in our discussion of the Incompleteness Theorems. Of course, since "S" is intended to stand for the successor operation, these terms can be thought of as denoting the natural numbers 0, 1, 2, 3, ... In general, for each natural number n there is a term denoting n that consists of n "S"s followed by "0". We will write "S^n0" as an abbreviation for this term, and call it the *numeral* for n.[2]

Formulas are, intuitively, expressions that make assertions about natural numbers. Like the definition of terms, the definition of formulas requires the use of parentheses in formulas to avoid ambiguity in their structure; again, the definition does not make any reference to meanings.

using (in our metalanguage) to denote expressions in the language of PA. When we speak, for example, of "the expression $(t + u)$," we do not mean the expression consisting of the five symbols "(", "t", "$+$", "u", and ")"; we mean the result of putting the symbol "$+$" between the expressions denoted by "t" and "u", and then surrounding the result with parentheses. The resulting expression may consist of more than five symbols, and its second symbol is not the letter "t" but, rather, the first symbol in the expression denoted by "t".

[2] It is important to understand that, in introducing this notation, we are not changing our definition of the language of PA. For example, the expression "S^40" is not a term in the language of PA. It is an abbreviation in our metalanguage that denotes the expression "$SSSS0$", and it is only this latter expression that is actually a term in the language of PA.

DEFINITION 7.2. The *formulas* of the language of PA are the expressions generated by the following rules:

   (i)   If t and u are terms, then $(t = u)$ and $(t < u)$ are formulas. Formulas of these kinds are called *atomic formulas*.

  (ii) If P and Q are formulas, then so are $\neg P, (P \wedge Q), (P \vee Q)$, $(P \rightarrow Q)$, and $(P \leftrightarrow Q)$.

 (iii) If P is a formula, then for each variable $x_i, \forall x_i P$ and $\exists x_i P$ are formulas.

 (iv) Nothing else is a formula.

For example, by rule (i), "$(x_1 = x_2)$" and "$((x_1 + x_3) = (x_2 + x_3))$" are formulas. Thus, by rules (ii) and (iii), "$(((x_1 + x_3) = (x_2 + x_3)) \rightarrow (x_1 = x_2))$" and "$\forall x_1 \forall x_2 \forall x_3 (((x_1 + x_3) = (x_2 + x_3)) \rightarrow (x_1 = x_2))$" are formulas. Of course, this last formula expresses the cancellation law for addition.

The distinction between *free* and *bound* occurrences of variables in formulas can also be defined without reference to the meanings of the symbols, as follows:

DEFINITION 7.3. All occurrences of variables in atomic formulas are free. When formulas are combined using logical connectives, the free or bound nature of occurrences of variables in them is unchanged. In formulas of the form $\forall x_i P$ or $\exists x_i P$, the free or bound nature of variables is the same as in P, except that all occurrences of x_i are bound. A formula in which all occurrences of variables are bound is called a *sentence*.

If P is a formula in which the only variables occurring free are $x_1, x_2, \ldots, x_n$, then we sometimes write it as $P(x_1, x_2, \ldots, x_n)$ to stress this fact. If $t_1, t_2, \ldots, t_n$ are terms, then $P(t_1, t_2, \ldots, t_n)$ is the formula that results from replacing all free occurrences of x_1 by t_1, all free occurrences of x_2 by t_2, and so on. We will be most interested in such substitutions when the terms $t_1, t_2, \ldots, t_n$ are numerals. For example, let $P(x_1, x_2)$ be the formula "$\exists x_3 ((x_1 + x_3) = x_2)$", which says that some natural number can be added to x_1 to get x_2. Then $P(S^2 0, S^3 0)$ is the sentence "$\exists x_3 ((SS0 + x_3) = SSS0)$", which makes the true assertion that some natural number can be added to 2 to get 3.

We are now ready to list the axioms of PA.

DEFINITION 7.4. The following sentences are nonlogical axioms of PA. (Axioms (i), (ii), and (viii) correspond to the Peano Postulates as defined in Chapter 2, and axioms (iii)–(vii) say that the symbols "+", "·", and "<" satisfy the appropriate definitions.)

(i) $\forall x_1 \neg (Sx_1 = 0)$.
(ii) $\forall x_1 \forall x_2 ((Sx_1 = Sx_2) \rightarrow (x_1 = x_2))$.
(iii) $\forall x_1 ((x_1 + 0) = x_1)$.
(iv) $\forall x_1 \forall x_2 ((x_1 + Sx_2) = S(x_1 + x_2))$.
(v) $\forall x_1 ((x_1 \cdot 0) = 0)$.
(vi) $\forall x_1 \forall x_2 ((x_1 \cdot Sx_2) = ((x_1 \cdot x_2) + x_1))$.
(vii) $\forall x_1 \forall x_2 ((x_1 < x_2) \leftrightarrow \exists x_3 ((x_1 + Sx_3) = x_2))$.

In addition, for each formula $P(x_1, x_2, \ldots, x_n)$, the following sentence is a nonlogical axiom of PA:

(viii) $\forall x_2 \forall x_3 \ldots \forall x_n ((P(0, x_2, \ldots, x_n) \wedge \forall x_1 (P(x_1, x_2, \ldots, x_n) \rightarrow P(Sx_1, x_2, \ldots, x_n))) \rightarrow \forall x_1 P(x_1, x_2, \ldots, x_n))$.

Note that axiom (viii), which expresses the principle of mathematical induction, is an axiom scheme; in other words, it is actually an infinite collection of axioms, one for each formula $P(x_1, x_2, \ldots, x_n)$.

In addition to these nonlogical axioms, the formal system PA also includes logical axioms and rules of inference for first-order logic. A list of such axioms and rules can be found in any textbook on logic. Different logic textbooks sometimes have slightly different sets of axioms and rules, but these differences will not be important to us in this chapter; the Incompleteness Theorems can be proven using the axioms and rules from any logic textbook. However, for the purpose of giving examples in this chapter it may be helpful to mention a few specific rules. Many logic textbooks use a rule called *universal elimination* (or sometimes *universal instantiation*) that allows us to infer $P(t)$ from $\forall x_i P(x_i)$, as long as t is a term that is substitutable[3] for x_i in $P(x_i)$. A second rule called the *substitution* rule (or sometimes, the *Law of the Indiscernability of Identicals*) says that if t and u are substitutable for x_i in $P(x_i)$, then from $(t = u)$ and $P(t)$ we may infer $P(u)$.

Sometimes in this chapter we will also be interested in extensions of PA. We will say that a formal system is an *extension* of PA if it is exactly the same as PA, except that it may include additional nonlogical axioms. Note that we consider PA itself to be an extension of PA; it is the extension in which no nonlogical axioms are added. All other extensions are formed by adding more nonlogical axioms to those listed in Definition 7.4.

DEFINITION 7.5.   Suppose that T is an extension of PA. A *proof* of a sentence Q in T is a sequence $P_1, P_2, \ldots, P_n$ of formulas such that, for

---

[3]   For the definition of substitutability, and a discussion of the reason for this restriction, see Enderton (2001, 113). We will mostly be interested in using this rule when t is a numeral, and numerals are substitutable for all variables in all formulas.

$i = 1, 2, \ldots, n$, either P_i is an axiom of T or P_i follows from earlier formulas in the proof by one of the rules of inference of T, and $P_n = Q$. If there is a proof of Q in T, then we say that Q is a *theorem* of T, and write $T \vdash Q$.

DEFINITION 7.6.   Suppose that T is an extension of PA. If there is a sentence P such that both P and $\neg P$ are theorems of T, then we say that T is *inconsistent*. If not, then it is *consistent*. If there is a sentence P such that neither P nor $\neg P$ is a theorem of T, then we say that T is *incomplete*. If not, then it is *complete*.

Here is an example of a proof of the sentence $((S^3 0 + S^2 0) = S^5 0)$ in PA. Next to each step in the proof, we have listed the justification for the step – either the fact that it is an axiom of PA, or the previous steps from which it follows and the rule of inference that is being used. In this proof we have used the rules of universal elimination and substitution mentioned earlier; only minor changes would be needed to make this proof conform to the logical axioms and rules of any logic textbook. The numbering of axioms here (and later in this chapter) refers to the list of axioms in Definition 7.4:

(1)
1. $\forall x_1((x_1 + 0) = x_1)$ — Axiom (iii)
2. $((SSS0 + 0) = SSS0)$ — 1, Universal elimination
3. $\forall x_1 \forall x_2((x_1 + Sx_2) = S(x_1 + x_2))$ — Axiom (iv)
4. $\forall x_2((SSS0 + Sx_2) = S(SSS0 + x_2))$ — 3, Universal elimination
5. $((SSS0 + S0) = S(SSS0 + 0))$ — 4, Universal elimination
6. $((SSS0 + S0) = SSSS0)$ — 2, 5, Substitution
7. $((SSS0 + SS0) = S(SSS0 + S0))$ — 4, Universal elimination
8. $((SSS0 + SS0) = SSSSS0)$ — 6, 7, Substitution

Having found this proof, we can say that PA $\vdash ((S^3 0 + S^2 0) = S^5 0)$.

The reader may have noticed that the example we have given here is the same as the example we used in Chapter 3 to illustrate the definition of addition of natural numbers, Definition 3.10. In fact, the reasoning in our proof of the sentence $((S^3 0 + S^2 0) = S^5 0)$ is just a formal version of the informal reasoning we used in (19) of Chapter 3 to show that $3 + 2 = 5$. In general, informal reasoning showing that some sentence Q follows from the axioms of PA can be turned into a formal proof of Q in PA, but we will usually not go to the trouble of performing this transformation. In the future, if we are able to show informally that Q follows from the axioms of PA, then we will conclude that PA $\vdash Q$. Readers experienced in formal logic should be able to turn our informal arguments into formal proofs. More precisely, and more generally, for any extension T of PA and any sentences $P_1, P_2, \ldots, P_n$ and Q, if $T \vdash P_1, T \vdash P_2, \ldots, T \vdash P_n$, and Q is a logical consequence of $P_1, P_2, \ldots, P_n$, then we can conclude that $T \vdash Q$. We will refer to this principle as the *closure of T under logical consequence*, and we will use it often in the rest of this chapter.

Of course, there is nothing special about the equation $3 + 2 = 5$. For any natural numbers a, b, and c, if $a + b = c$ then a proof similar to the one given earlier can be used to show that PA $\vdash ((S^a 0 + S^b 0) = S^c 0)$. Also, if $a + b \neq c$ then it can be shown that PA $\vdash \neg((S^a 0 + S^b 0) = S^c 0)$. For example, let us see why PA $\vdash \neg((S^3 0 + S^2 0) = S^4 0)$. From $((S^3 0 + S^2 0) = S^4 0)$, together with the equation $((S^3 0 + S^2 0) = S^5 0)$, which we have already proven, we could infer $(S^5 0 = S^4 0)$. But then, by axiom (ii), it follows that $(S^4 0 = S^3 0)$. Applying axiom (ii) again we can conclude $(S^3 0 = S^2 0)$, and applying it two more times we finally reach the conclusion $(S0 = 0)$. But this contradicts axiom (i). Thus the assumption that $((S^3 0 + S^2 0) = S^4 0)$ has led to a contradiction, so we can conclude $\neg((S^3 0 + S^2 0) = S^4 0)$. This informal reasoning shows that our theorem $((S^3 0 + S^2 0) = S^5 0)$ and axioms (i) and (ii) of PA logically imply $\neg((S^3 0 + S^2 0) = S^4 0)$. By the closure of PA under logical consequence, this informal reasoning can be turned into a formal proof showing that PA $\vdash \neg((S^3 0 + S^2 0) = S^4 0)$. We can summarize our conclusions as follows:

THEOREM 7.7.   *Let $P(x_1, x_2, x_3)$ be the formula* "$((x_1 + x_2) = x_3)$". *Then, for any natural numbers a, b, and c, if $a + b = c$ then* PA $\vdash P(S^a 0, S^b 0, S^c 0)$, *and if $a + b \neq c$ then* PA $\vdash \neg P(S^a 0, S^b 0, S^c 0)$.

This is our first example of an idea that will play a central role in the proofs of the Incompleteness Theorems. Here is the general definition:

DEFINITION 7.8.   For any positive integer k, let $\mathbb{N}^k$ be the set of all sequences of natural numbers of length k. We will use the notation $\langle a_1, a_2, \ldots, a_k \rangle$ to denote an element of $\mathbb{N}^k$. A set $A \subseteq \mathbb{N}^k$ is called *representable* if there is a formula $P(x_1, x_2, \ldots, x_k)$ such that, for every sequence $\langle a_1, a_2, \ldots, a_k \rangle \in \mathbb{N}^k$:
   (i) if $\langle a_1, a_2, \ldots, a_k \rangle \in A$, then PA $\vdash P(S^{a_1} 0, S^{a_2} 0, \ldots, S^{a_k} 0)$; and
   (ii) if $\langle a_1, a_2, \ldots, a_k \rangle \notin A$, then PA $\vdash \neg P(S^{a_1} 0, S^{a_2} 0, \ldots, S^{a_k} 0)$.
In this case, we say that the formula P *represents* the set A.

Intuitively, if P represents A, then we can think of $P(x_1, x_2, \ldots, x_k)$ as saying that $\langle x_1, x_2, \ldots, x_k \rangle$ is an element of A. Thus, for any particular sequence of numbers $\langle a_1, a_2, \ldots, a_k \rangle$, $P(S^{a_1}0, S^{a_2}0, \ldots, S^{a_k}0)$ is a sentence in the language of PA that says that $\langle a_1, a_2, \ldots, a_k \rangle$ is an element of A, and $\neg P(S^{a_1}0, S^{a_2}0, \ldots, S^{a_k}0)$ is a sentence saying that $\langle a_1, a_2, \ldots, a_k \rangle$ is not an element of A. Furthermore, the definition of representable requires that facts about whether or not particular sequences are elements of A, when expressed in the language of PA in this way using P, must always be *provable* in PA.

Using this terminology, we can rephrase Theorem 7.7 by saying that the formula "$((x_1 + x_2) = x_3)$" represents the set $\{\langle a, b, c \rangle \in \mathbb{N}^3 : a + b = c\}$. Similarly, it is not hard to show that the formula "$((x_1 \cdot x_2) = x_3)$" represents the set $\{\langle a, b, c \rangle \in \mathbb{N}^3 : a \cdot b = c\}$ (see exercise 2). For a more complicated example, let $P(x_1)$ be the formula "$((S0 < x_1) \wedge \neg \exists x_2 \exists x_3 (((x_2 \cdot x_3) = x_1) \wedge ((x_2 < x_1) \wedge (x_3 < x_1))))$". Intuitively, $P(x_1)$ says that x_1 is a number larger than 1 that cannot be written as a product of two smaller numbers; in other words, it says that x_1 is a prime number. It can be shown that, for every natural number n, if n is a prime number then PA $\vdash P(S^n 0)$, and if n is not a prime number then PA $\vdash \neg P(S^n 0)$. Thus, $P(x_1)$ represents the set of prime numbers.[4] The first step in proving the Incompleteness Theorems is to show that many more sets are representable.

A key idea that is often used in proving that sets are representable is the fact that it is possible to encode a finite sequence of natural numbers with a single natural number. To see how this can be done, first let p_n denote the nth prime number. Thus, $p_1 = 2$, $p_2 = 3$, $p_3 = 5$, and so on. Then, for any sequence of natural numbers $\langle a_1, a_2, \ldots, a_k \rangle$, we define the *code number* of the sequence to be the number

$$(2) \quad \#\langle a_1, a_2, \ldots, a_k \rangle = p_1^{a_1 + 1} \cdot p_2^{a_2 + 1} \cdots p_k^{a_k + 1}.$$

For example,

$$(3) \quad \#\langle 2, 3, 0 \rangle = p_1^{2+1} \cdot p_2^{3+1} \cdot p_3^{0+1} = 2^3 \cdot 3^4 \cdot 5^1 = 8 \cdot 81 \cdot 5 = 3,240;$$
$$\#\langle 2, 3 \rangle = p_1^{2+1} \cdot p_2^{3+1} = 2^3 \cdot 3^4 = 8 \cdot 81 = 648.$$

---

[4] There is a difference between a natural number n, which is an element of $\mathbb{N}$, and the sequence of length 1 whose only term is n, which would be written $\langle n \rangle$ and is an element of $\mathbb{N}^1$. Strictly speaking, Definition 7.6 says that $P(x_1)$ represents not the set of prime numbers but, rather, $\{\langle n \rangle \in \mathbb{N}^1 : n$ is a prime number$\}$. However, we will usually ignore this distinction and speak of a subset of $\mathbb{N}$ as being representable when the corresponding subset of $\mathbb{N}^1$ is.

The reason for including "+1" in the exponents in (2) was simply to make sure that the two sequences in (3) would be assigned different code numbers.

Using the fact that every positive integer can be written as a product of prime numbers in a unique way, it is not hard to show that this coding method always assigns different code numbers to different sequences. In fact, given any number s, we can use the prime factorization of s to determine whether or not s is the code number of a sequence, and if so what unique sequence has code number s. For example,

$$(4) \quad 73,500 = 2^2 \cdot 3^1 \cdot 5^3 \cdot 7^2 = p_1^{1+1} \cdot p_2^{0+1} \cdot p_3^{2+1} \cdot p_4^{1+1} = \#\langle 1, 0, 2, 1 \rangle.$$

On the other hand, $588 = 2^2 \cdot 3^1 \cdot 7^2 = p_1^{1+1} \cdot p_2^{0+1} \cdot p_4^{1+1}$ is not the code number of any sequence, because p_3 has been skipped in the prime factorization.

If s is a sequence code number, then we define $l(s)$ to be the length of the sequence encoded by s, and $(s)_i$ to be the ith term of the sequence. For example, by (4), $l(73,500) = 4$ and $(73,500)_3 = 2$. Our next theorem is rather difficult to prove, but it is one of the reasons why sequence numbers are so useful in showing that sets are representable.[5]

THEOREM 7.9. *The following sets are representable:*
  (i) $\{\langle s, n \rangle \in \mathbb{N}^2 : s$ *is a sequence code number and* $l(s) = n\}$.
  (ii) $\{\langle s, i, a \rangle \in \mathbb{N}^3 : s$ *is a sequence code number,* $1 \leq i \leq l(s)$, *and* $(s)_i = a\}$.

An example should make it clear why sequence code numbers are useful for showing that sets are representable. Recall that, for any natural number $n \geq 1$, the product $1 \cdot 2 \cdots n$ is called n *factorial*, and is denoted $n!$. We will show that the set $F = \{\langle n, m \rangle \in \mathbb{N}^2 : n \geq 1$ and $m = n!\}$ is representable. To do this, we must find a formula $P(x_1, x_2)$ that says that $x_1 \geq 1$ and $x_2 = x_1!$. But the factorial symbol is not in the language of PA and, more importantly, neither is the "$\cdots$" that we used in our explanation of the meaning of "factorial." How can we talk about the product of a list of numbers in the language of PA?

To solve this problem, note that in order to compute $n!$ we start with the number 1, and then multiply by 2, and then multiply by 3, and so on until we reach n. Thus, in the course of computing n! we

---

[5] A proof can be found in many logic textbooks. See, for example, Enderton (2001, 220–1).

will actually compute a whole list of intermediate results: 1, $1 \cdot 2 = 2$, $1 \cdot 2 \cdot 3 = 6, \ldots, 1 \cdot 2 \cdots n = n!$. This list of intermediate results is just the sequence of all factorials up to n, and it can be encoded with a single code number s:

(5) $\quad s = \#\langle 1!, 2!, \ldots, n! \rangle$.

We will represent the set F with a formula $P(x_1, x_2)$ that talks about this sequence by talking about the code number s. Note that the terms in the sequence can be specified by a recursive definition: $(s)_1 = 1$, and if $1 \leq i < n$ then $(s)_{i+1} = (i+1) \cdot (s)_i$.

Let $P(x_1, x_2)$ be the following formula:

(6) $\quad ((0 < x_1) \wedge \exists s(s$ is a sequence code number $\wedge \; l(s) = x_1 \wedge$
$\qquad (s)_1 = 1 \wedge \forall i(1 \leq i < x_1 \rightarrow (s)_{i+1} = (i+1) \cdot (s)_i) \wedge (s)_{x_1} = x_2))$.

Of course, (6) is not really a formula in the language of PA, but it can be turned into one. To say that s is a sequence code number and $l(s) = x_1$, we can use the formula that represents the set in part (i) of Theorem 7.9. Similarly, by part (ii) of Theorem 7.9, all other parts of (6) can be expressed by formulas in the language of PA. We must also replace the bound variables s and i in (6) by variables in the language of PA, such as x_3 and x_4, and we must add some parentheses. When we say that $P(x_1, x_2)$ is the formula (6), what we mean is that it is the result of translating (6) into the language of PA in this way. It is now tedious, but not really very difficult, to prove that for all natural numbers n and m, if $n \geq 1$ and $m = n!$ then PA $\vdash P(S^n 0, S^m 0)$, and if not then PA $\vdash \neg P(S^n 0, S^m 0)$.

This example illustrates a method that can often be used to show that a set $A \subseteq \mathbb{N}^k$ is representable. The method can be used any time there is a step-by-step procedure for determining whether or not a given sequence $\langle a_1, a_2, \ldots, a_k \rangle$ is an element of A. The procedure must consist of a finite list of precise instructions, and the process of following these instructions must be completely mechanical, requiring no originality or use of random devices such as the flipping of a coin. Furthermore, any answer produced by the procedure must be produced in a finite number of steps. Procedures with these properties are known as *algorithms*. If there is an algorithm for determining whether or not a given sequence is an element of A, then we say that A is *decidable*. By imitating the method used in the last example, it can be shown that every decidable set is representable.

THEOREM 7.10.   *If a set $A \subseteq \mathbb{N}^k$ is decidable, then it is representable.*

Idea of Proof. It is not possible to give a rigorous proof of this theorem here, since we have not given a precise definition of "algorithm."[6] Nevertheless, it may give the reader a better feeling for representable sets to see at least roughly why the theorem is true.

Suppose that A is decidable. Then there is an algorithm that can be used to determine whether or not a given sequence $\langle a_1, a_2, \ldots, a_k \rangle$ is an element of A. As in our last example, it is possible to construct a sequence of natural numbers that encodes all of the steps in carrying out this algorithm, and this sequence can then be encoded with a single sequence code number s. We can then represent A with a formula in the language of PA that talks about this number s. The formula that represents A will say something like this:

(7) $\exists s(s$ is a sequence code number that encodes the steps in
       executing the algorithm for determining whether or not
       $\langle x_1, x_2, \ldots, x_k \rangle \in A$, and this execution of the algorithm
       determines that $\langle x_1, x_2, \ldots, x_k \rangle \in A$).   ∎

In fact, the converse of Theorem 7.10 is true as well: If a set is representable, then it is decidable (see exercise 10). Thus, the representable sets are precisely the same as the decidable sets. However, we will not need this fact in the proofs of the Incompleteness Theorems.

We have seen that it is useful to assign code numbers to sequences of natural numbers. One way of thinking about the reason for this is that, by introducing the idea of sequence numbers, we have in effect expanded the expressive power of the language of PA. We have not changed the definition of the language of PA, but we have shown that this language can be used to talk about more than just natural numbers; it can also be used to talk indirectly about sequences of natural numbers, by talking about the code numbers for these sequences. Similarly, we will be able to use the language of PA to talk about other objects if we can find a way to assign code numbers to these objects. The idea behind the proofs of the Incompleteness Theorems is to use the language of PA to talk about the formal system PA itself! In fact, we will show that many facts about PA are, when properly encoded as sentences in the language of PA, provable in PA. To do this, we must assign numbers to expressions in the language of PA. We begin by assigning a number to each symbol in the language of PA:

---

[6]   For a precise definition, see Cutland (1980).

Symbol	Number
∧	0
∨	1
¬	2
→	3
↔	4
∀	5
∃	6
(	7
)	8
=	9
0	10
S	11
<	12
+	13
·	14
x_1	15
x_2	16
⋮	⋮
x_n	$14 + n$
⋮	⋮

An expression in the language of PA is a sequence of symbols. Replacing each symbol by the corresponding number gives us a sequence of natural numbers, and we already know how this sequence can be encoded with a single code number. This code number is called the *Gödel number* of the original expression.

For example, let P be the formula "$(0 < S0)$". The first symbol of P is "(", which corresponds to the number 7. Next is "0", whose number is 10. Continuing to list the numbers corresponding to the symbols in P, we get the sequence $\langle 7, 10, 12, 11, 10, 8 \rangle$. The Gödel number of P, denoted #P, is the code number of this sequence:

(8) $\#P = \#\langle 7, 10, 12, 11, 10, 8 \rangle = 2^8 \cdot 3^{11} \cdot 5^{13} \cdot 7^{12} \cdot 11^{11} \cdot 13^9$
$= 2,318,302,007,414,142,079,698,247,490,754,875,$
$294,062,500,000,000.$

As this example shows, the Gödel number of even a very simple expression in the language of PA can be extremely large!

Let us try an example working in the opposite order. Is the number 528,958,107,648 the Gödel number of an expression in the language of PA? Factoring this number into primes, we find that:

(9) $528,958,107,648 = 2^{12} \cdot 3^{17} = \#\langle 11, 16 \rangle.$

Now 11 is the number corresponding to the symbol "S", and 16 corresponds to "x_2". Therefore 528,958,107,648 is the Gödel number of the term "Sx_2".

Now that we have assigned numbers to the expressions in the language of PA, a number of new representability questions naturally suggest themselves. For example, some natural numbers (such as 528,958, 107,648) are Gödel numbers of terms in the language of PA, and some are not. Is the set of Gödel numbers of terms representable? According to Theorem 7.10, this set is representable if it is decidable, so we are led to ask whether or not there is an algorithm that, given a natural number n, will determine whether or not n is the Gödel number of a term. In fact, it seems likely that there is. By factoring n into primes, we can determine whether or not it is the Gödel number of some expression in the language of PA, and if so what that expression is. Now recall our earlier observation that the definition of terms makes no reference to the meanings of the symbols, but talks only about how the symbols are arranged. As a result, to determine whether or not an expression is a term we need only check whether or not the symbols are arranged in accordance with the rules in Definition 7.1, and this checking can be carried out in a finite number of steps. This leads us to our next theorem.

THEOREM 7.11.   *The following sets are representable*:
  (i) $\{n \in \mathbb{N} : n$ *is the Gödel number of a term*$\}$.
  (ii) $\{n \in \mathbb{N} : n$ *is the Gödel number of a formula*$\}$.
  (iii) $\{n \in \mathbb{N} : n$ *is the Gödel number of a sentence*$\}$.

Idea of Proof. One way to prove that the set in (i) is representable is to use the ideas of the previous paragraph to show that the set is decidable, and then apply Theorem 7.10. However, the reader might find it enlightening to see how one might go about writing a formula $P(x_1)$ that represents this set. According to the proof of Theorem 7.10, one way to do this is to write a formula that talks about a sequence code number s that encodes the information involved in checking whether or not a given number is the Gödel number of a term. What information should be encoded in s?

The most complicated part of Definition 7.1 is clause (iii), which says that in order to determine whether or not an expression v is a term, we must check whether or not it is a certain kind of combination of simpler expressions t and u that are terms. An important (and easily proven) fact about such combinations is that the Gödel number of the expression v is larger than the Gödel numbers of the simpler expressions t and u from which it is constructed. It follows that in order to determine whether or not n is the Gödel number of a term, it may be necessary to check whether certain smaller numbers are Gödel numbers of terms. This suggests that the number s should encode information not only about whether or not n is the Gödel number of a term, but also about whether or not smaller numbers are Gödel numbers of terms.

Here is one way of defining such a sequence code number s. For any natural number n, let $\langle a_1, a_2, \ldots, a_n \rangle$ be the sequence defined as follows:

$$(10) \quad a_i = \begin{cases} 1, & \text{if } i \text{ is the Godel number of a term;} \\ 0, & \text{if not.} \end{cases}$$

Let $s = \#\langle a_1, a_2, \ldots, a_n \rangle$. We will say that s *checks Gödel numbers of terms up to* n. Then the formula $P(x_1)$ that represents the set of Gödel numbers of terms will be

$$(11) \quad \exists s(s \text{ checks Gödel numbers of terms up to } x_1 \land (s)_{x_1} = 1).$$

Of course, we must still explain how to express "s checks Gödel numbers of terms up to x_1" in the language of PA. This turns out to be tedious but not really very difficult. According to (10) and Definition 7.1, we must say that s is a sequence code number, $l(s) = x_1$, and for every i from 1 to x_1, $(s)_i$ is either 0 or 1, and $(s)_i = 1$ if and only if either i is the Gödel number of the expression "0", or it is the Gödel number of the expression "x_j" for some j, or it is the Gödel number of an expression constructed in a certain way from expressions with smaller Gödel numbers j and k and $(s)_j = (s)_k = 1$. It can be shown that all of this can be expressed in the language of PA, and the resulting formula $P(x_1)$ represents the set of Gödel numbers of terms.

The proofs for parts (ii) and (iii) are similar, using Definitions 7.2 and 7.3 instead of Definition 7.1. ■

It is interesting to note that Theorem 7.11 can be thought of as saying that the early parts of our description of PA in this chapter – namely, Definitions 7.1, 7.2, and 7.3 – can be expressed in the language of PA, once the objects involved have been encoded with Gödel numbers. Let us see how far we can take this idea. Definition 7.4 defines the nonlogical axioms of PA, and it can be used to give an algorithm for determining, for any given sentence, whether or not that sentence is a nonlogical axiom of PA. It follows, by Theorem 7.10, that $\{n : n$ is the Gödel number of a nonlogical axiom of PA$\}$ is representable. If T is an extension of PA, then

$\{n : n$ is the Gödel number of a nonlogical axiom of $T\}$ may not be representable. However, using Theorem 7.10 we can show that if there is an algorithm for determining, for any given sentence, whether or not that sentence is a nonlogical axiom of T, then the set of Gödel numbers of nonlogical axioms of T will be representable. In this case we will call T a *decidable* extension of PA.

Definition 7.5 defines the proofs and theorems of extensions of PA, but it cannot be expressed in the language of PA until we have assigned Gödel numbers to proofs. Fortunately, this is not difficult to do. Suppose $P_1, P_2, \ldots, P_n$ is a proof. To encode this proof with a single number, first compute the Gödel numbers of the formulas in the proof to get the sequence of numbers $\langle \#P_1, \#P_2, \ldots, \#P_n \rangle$. Then, encode this sequence of numbers with a single number. The result is the number $\#\langle \#P_1, \#P_2, \ldots, \#P_n \rangle$, which we will call the *Gödel number* of the proof.

For example, here is a two-line proof in PA:

$$(12) \quad \begin{array}{lll} 1. & \forall x_1 \neg(Sx_1 = 0) & \text{Axiom (i)} \\ 2. & \neg(S0 = 0) & \text{1, Universal elimination} \end{array}$$

To compute the Gödel number of this proof, we first compute the Gödel numbers of the individual formulas in the proof. We will call these numbers a and b.

$$
\begin{aligned}
(13) \quad a &= \#\forall x_1 \neg(Sx_1 = 0) \\
&= \#\langle 5, 15, 2, 7, 11, 15, 9, 10, 8 \rangle \\
&= 2^6 \cdot 3^{16} \cdot 5^3 \cdot 7^8 \cdot 11^{12} \cdot 13^{16} \cdot 17^{10} \cdot 19^{11} \cdot 23^9 \\
&= 1,753,676,934,331,034,663,878,331,554,836,419,381, \\
&\qquad 406,955,754,926,957,007,067,649,127,704,826,043, \\
&\qquad 281,843,299,944,000; \\
b &= \#\neg(S0 = 0) \\
&= \#\langle 2, 7, 11, 10, 9, 10, 8 \rangle \\
&= 2^3 \cdot 3^8 \cdot 5^{12} \cdot 7^{11} \cdot 11^{10} \cdot 13^{11} \cdot 17^9 \\
&= 139,676,240,233,639,528,072,573,470,580,523,995, \\
&\qquad 351,501,719,427,734,375,000.
\end{aligned}
$$

Although these two numbers are extremely large, they are small enough that we can write them down, with each requiring fewer than 100 digits. But they are miniscule in comparison to the Gödel number of the proof. The Gödel number of the proof is the code number of the sequence consisting of a and b:

(14)   Gödel number of proof $= \#\langle a, b \rangle = 2^{a+1} \cdot 3^{b+1}$.

This number is so large that it would require about 5.28×10^{86} digits to write it down. All the paper in the world would not suffice to write down the Gödel number of this simple two-line proof. Nevertheless, it is a well-defined, finite natural number.

Now let us return to the task of expressing Definition 7.5 in the language of PA. The second sentence of Definition 7.5 defines the proofs of any extension T of PA, and we can use it to try to construct an algorithm for determining whether or not a given sequence of formulas is a proof of a given sentence in T. Note that one of the things this algorithm must do is to check whether or not a formula in the sequence is an axiom of T, and this cannot be done unless T is a decidable extension of PA. But if T is decidable, then such an algorithm can be constructed. (In fact, some logic textbooks come with software that implements such an algorithm for some formal system, although the formal system used is usually not an extension of PA.) Applying Theorem 7.10 now leads to our most important representability result:

THEOREM 7.12.   *Suppose that T is a decidable extension of PA. Then $\{\langle n, m \rangle \in \mathbb{N}^2 : n$ is the Gödel number of a sentence and m is the Gödel number of a proof of that sentence in $T\}$ is representable.*

Since Theorem 7.12 is so important, it may be useful to spell out more fully what it means. Let T be a decidable extension of PA. According to the definition of representability, what the theorem says is that there is some formula $\text{Proof}_T(x_1, x_2)$ in the language of PA such that if n is the Gödel number of a sentence and m is the Gödel number of a proof of that sentence in T, then PA $\vdash$ $\text{Proof}_T(S^n 0, S^m 0)$, and if not, then PA $\vdash \neg \text{Proof}_T(S^n 0, S^m 0)$. We can think of $\text{Proof}_T(x_1, x_2)$ as expressing in the language of PA the statement that x_1 is the Gödel number of a sentence and x_2 is the Gödel number of a proof of that sentence in T. In fact, we can construct the formula $\text{Proof}_T(x_1, x_2)$ simply by translating the second sentence of Definition 7.5 into the language of PA. The result is a formula that says something like this:

(15) x_1 is the Gödel number of a sentence $\wedge\, x_2$ is a sequence code number $\wedge\, \forall i (1 \leq i \leq l(x_2) \rightarrow ((x_2)_i$ is the Gödel number of a non-logical axiom of $T \vee (x_2)_i$ is the Gödel number of a logical axiom $\vee\, (x_2)_i$ is the Gödel number of a formula that follows by a logical rule of inference from formulas whose Gödel numbers appear earlier in the sequence coded by $x_2)) \wedge (x_2)_{l(x_2)} = x_1$.

The third sentence of Definition 7.5 defines the theorems of T. It says that a sentence is a theorem of T if it has a proof, and it is easy to translate this into the language of PA. Let $\text{Theorem}_T(x_1)$ be the formula

(16) $\exists x_2 \text{Proof}_T(x_1, x_2)$.

Then $\text{Theorem}_T(x_1)$ says that x_1 is the Gödel number of a sentence, and there is a number that is the Gödel number of a proof of that sentence in T. In other words, it says that x_1 is the Gödel number of a theorem of T.

It is natural to wonder at this point whether or not the formula $\text{Theorem}_T(x_1)$ represents the set of Gödel numbers of theorems of T. In other words, is it the case that if n is the Gödel number of a theorem of T then PA $\vdash \text{Theorem}_T(S^n0)$, and if not then PA $\vdash \neg\text{Theorem}_T(S^n0)$? Let us see if we can prove these statements. Suppose first that n is the Gödel number of a sentence P that is a theorem of T. Then there is a proof of P in T. Let m be the Gödel number of this proof. Then according to our comments after Theorem 7.12, PA $\vdash \text{Proof}_T(S^n0, S^m0)$. But according to (16), $\text{Theorem}_T(S^n0)$ is the sentence $\exists x_2 \text{Proof}_T(S^n0, x_2)$, and this follows logically from $\text{Proof}_T(S^n0, S^m0)$. Therefore, by the closure of PA under logical consequence, PA $\vdash \text{Theorem}_T(S^n0)$.

Now suppose that n is not the Gödel number of a theorem of T. For example, suppose that n is the Gödel number of a sentence P, but P is not a theorem of T. Then no number can be the Gödel number of a proof of P in T. It follows that, for every natural number m, PA $\vdash \neg\text{Proof}_T(S^n0, S^m0)$. In other words, for every m there is a proof in PA of the sentence $\neg\text{Proof}_T(S^n0, S^m0)$. But these proofs may all be very different, because the reasons why different numbers fail to be Gödel numbers of proofs of P may be different. There may be no uniform reason why no number can be the Gödel number of a proof of P, in which case there would be no proof in PA of the single sentence $\forall x_2 \neg\text{Proof}_T(S^n0, x_2)$. But this sentence is equivalent to $\neg\text{Theorem}_T(S^n0)$, so it appears that there may be no proof of $\neg\text{Theorem}_T(S^n0)$ in PA. Thus, we have succeeded in proving the first half of the definition of representability – what we might call the "positive" half – but not the second, "negative," half.

Note that we are not questioning the accuracy of (16) as a translation of the definition of the word "theorem." The formula $\text{Theorem}_T(x_1)$, as defined by (16), does accurately express in the language of PA the definition of "theorem" from Definition 7.5. Thus, if n is the Gödel number of a theorem of T, then the sentence $\text{Theorem}(S^n0)$ is true, and if n is not the Gödel number of a theorem of T, then the sentence $\neg\text{Theorem}_T(S^n0)$ is true. The point we are making is that in the latter situation the sentence $\neg\text{Theorem}_T(S^n0)$, although true, may not be *provable* in PA. We will express this by saying that the formula Theorem$_T(x_1)$ *defines* the set of Gödel numbers of theorems of T, but we do not yet know if it *represents* that set.

Let us try another approach to showing that the set of Gödel numbers of theorems of T is representable. According to Theorem 7.10, we can show that this set is representable by showing that it is decidable, so let us see if we can come up with an algorithm for determining, for any given number n, whether or not n is the Gödel number of a theorem of T. We could begin by decoding the number n to determine whether or not it is the Gödel number of some sentence P. If it is, then we could search for a proof of P. For example, we might run through all possible proofs in order of increasing Gödel number, checking each to see if it is a correct proof of P. If P is a theorem then we will eventually find such a proof, and we will then know that P is a theorem. But if P is not a theorem, then this search will go on forever. Even very simple sentences sometimes have very complicated proofs, so there is no point in the search at which we can stop and be sure that if no proof has yet been found, then none will ever be found. Thus, if P is not a theorem then our proposed algorithm will run forever, never coming to a conclusion about whether or not P is a theorem. Once again, the "positive" half of the proof has worked but the "negative" half has failed.

By now you are probably beginning to suspect that the set of Gödel numbers of theorems of T is *not* representable – although, of course, our two failed attempts to prove that this set is representable do not constitute a proof that it is not! We will return to this issue later, but for now let us simply state formally the part of representability that we were able to prove:

THEOREM 7.13. *Suppose that T is a decidable extension of PA. If n is the Gödel number of a theorem of T, then* PA $\vdash$ Theorem$_T(S^n0)$.

Here is another way of stating this result. Let T be a decidable extension of PA, and suppose that $T \vdash P$. Let $n = \#P$. Then n is the Gödel number of a theorem of T, so by Theorem 7.13, PA $\vdash$ Theorem$_T(S^n0)$. Filling in the definition of n, this means that PA $\vdash$ Theorem$_T(S^{\#P}0)$. Thus, we have the following corollary of Theorem 7.13:

COROLLARY 7.14. *Suppose that T is a decidable extension of PA. If $T \vdash P$, then* PA $\vdash$ Theorem$_T(S^{\#P}0)$.

The numeral $S^{\#P}0$ will come up often in the rest of this chapter, so it may be useful to have an abbreviation for it. If P is any expression in the language of PA, then we will write $\ulcorner P \urcorner$ as an abbreviation for $S^{\#P}0$. In other words, $\ulcorner P \urcorner$ is the numeral for the Gödel number of P. In this notation, Corollary 7.14 says that if $T \vdash P$ then PA $\vdash$ Theorem$_T(\ulcorner P \urcorner)$.

There is just one more fact we need to discuss before we can prove the Incompleteness Theorems:

LEMMA 7.15 (Fixed Point Lemma). *Let $P(x_1)$ be any formula in the language of PA. Then there is a sentence Q such that*:

$$PA \vdash Q \leftrightarrow P(\ulcorner Q \urcorner).$$

The proof of the Fixed Point Lemma is beautiful and clever, but it is also somewhat technical, and the technicalities are not relevant to the proofs of the Incompleteness Theorems. We will therefore put off the proof until after our discussion of the Incompleteness Theorems.

The Fixed Point Lemma says that there is a sentence Q such that we can prove in PA that Q is equivalent to a statement about Q's own Gödel number. More informally, we might say that Q says something about its own Gödel number. What does Q say about its own Gödel number? Since the lemma can be applied to *any* formula $P(x_1)$, the answer is that we can find a sentence Q that says *whatever we want* about its own Gödel number! The key to proving the Incompleteness Theorems is to choose carefully what we want a sentence to say about its own Gödel number. In other words, we must make a careful choice of the formula to use as $P(x_1)$ in the Fixed Point Lemma.

Let us see what happens if we let $P(x_1)$ be the formula $\neg$Theorem$_T(x_1)$, for some decidable extension T of PA. Then, according to the Fixed Point Lemma, there must be a sentence G_T (the "Gödel sentence" for T) such that

(17) $PA \vdash G_T \leftrightarrow \neg$Theorem$_T(\ulcorner G_T \urcorner)$.

Informally, we might say that G_T says "My Gödel number is not the Gödel number of a theorem of T" or, in other words, "I am not a theorem of T."

Is G_T a theorem of T? If $T \vdash G_T$, then by Corollary 7.14, PA $\vdash$ Theorem$_T(\ulcorner G_T \urcorner)$. Therefore, by (17) and the closure of PA under logical consequence, PA $\vdash \neg G_T$. Note that, since T is an extension of PA, every proof in PA is also a proof in T. Therefore $T \vdash \neg G_T$, which, together with the fact that $T \vdash G_T$, implies that T is inconsistent! Thus, if T is consistent, then $T \nvdash G_T$; that is, G_T is not a theorem of T.

Is it possible that $T \vdash \neg G_T$? If so, then using (17) and the fact that T is an extension of PA we can conclude that $T \vdash \text{Theorem}_T(\ulcorner G_T \urcorner)$, and since $\text{Theorem}_T(x_1)$ is defined by (16), this means that

(18) $T \vdash \exists x_2 \text{Proof}_T(\ulcorner G_T \urcorner, x_2)$.

But, assuming again that T is consistent, we know that G_T is not a theorem of T, so no natural number m can be the Gödel number of a proof of G_T in T. Therefore, for every natural number m, $\text{PA} \vdash \neg \text{Proof}_T(\ulcorner G_T \urcorner, S^m 0)$, so since T is an extension of PA,

(19) for every number m, $T \vdash \neg \text{Proof}_T(\ulcorner G_T \urcorner, S^m 0)$.

The combination of (18) and (19) is not an outright inconsistency in T, because we do not have a single sentence P for which both P and $\neg P$ are theorems of T. Nevertheless, there is a kind of conflict between (18) and (19). According to (18), we can prove in T that there is a number with a certain property, but (19) says that we can prove, for each number m, that m does not have that property! We need a name for this kind of conflict.

DEFINITION 7.16.   We will say that an extension T of PA is *ω-inconsistent* if there is a formula $P(x_1)$ such that:
   (i) $T \vdash \exists x_1 P(x_1)$, and
   (ii) for every natural number m, $T \vdash \neg P(S^m 0)$.
If there is no such formula, then T is *ω-consistent*. Note that the sentences $\exists x_1 P(x_1), \neg P(0), \neg P(S0), \neg P(SS0), \ldots$ cannot all be true. Therefore, if all theorems of T are true, then T must be ω-consistent.

Rephrasing our conclusion from the last paragraph in this language, we can say that if T is consistent and $T \vdash \neg G_T$, then T is ω-inconsistent. In other words, if T is both consistent and ω-consistent, then $T \nvdash \neg G_T$. This last statement is actually slightly redundant; there is no need to include the assumption that T is consistent, because ω-consistency implies consistency:

LEMMA 7.17.   *If T is any extension of PA that is ω-consistent, then T is consistent.*

Proof. We use the fact that if T is inconsistent, then *every* sentence can be proven in T. A proof of this fact can be found in any logic textbook. It follows that if T is inconsistent then statements (i) and (ii) of Definition

7.16 are true (for *any* formula $P(x_1)$), so T is ω-inconsistent. In other words, if T is ω-consistent then it is consistent.   ∎

Putting together everything we have proven about the sentence G_T yields the first of Gödel's Incompleteness Theorems:

THEOREM 7.18 (First Incompleteness Theorem).   *Let T be a decidable extension of PA. Then there is a sentence G_T such that:*
  (i)   *If T is consistent then $T \not\vdash G_T$, and*
  (ii) *If T is ω-consistent then $T \not\vdash \neg G_T$.*

To understand the significance of the First Incompleteness Theorem, suppose that T is an extension of PA that is decidable and ω-consistent (and therefore also consistent, by Lemma 7.17). Then the First Incompleteness Theorem says that neither G_T nor $\neg G_T$ is a theorem of T. Thus, T is incomplete. (Of course, this is the reason for the name "Incompleteness Theorem.") Note that G_T is a sentence in the language of PA, which contains, in addition to the symbols of logic, symbols for talking about the operations of addition and multiplication, the successor operation, and the ordering of the natural numbers, and nothing else. Thus, G_T is simply a very complicated sentence about the arithmetic of the natural numbers. From the point of view of a classical mathematician, it is therefore either true or false, although of course an intuitionist might question this. In other words, from the classical point of view, the question "Is G_T true or false?" has an answer, but the formal system T is not powerful enough to determine this answer. Thus, the First Incompleteness Theorem says that no decidable, ω-consistent extension of PA can be powerful enough to determine the answers to all questions of number theory.

In fact, it is not really very difficult to determine the truth value of Gödel's sentence G_T for any decidable, consistent extension T of PA. Since G_T is not a theorem of T, there is no number that is the Gödel number of a proof of G_T in T. Therefore, the sentence $\exists x_2 \mathrm{Proof}_T(\ulcorner G_T \urcorner, x_2)$, which is simply $\mathrm{Theorem}_T(\ulcorner G_T \urcorner)$, is false. But by (17), the equivalence $G_T \leftrightarrow \neg \mathrm{Theorem}_T(\ulcorner G_T \urcorner)$ is a theorem of PA, and therefore true. It follows that G_T is true. To put the argument more informally: G_T says "I am not a theorem of T," and since it is not a theorem of T, what it says is true. Thus, G_T is a true sentence that is not provable in T.

In the last paragraph, we presented reasoning establishing that G_T is true. Does this not constitute a *proof* of G_T, contradicting our claim that G_T is unprovable? The answer is that we have not claimed that G_T is absolutely unprovable, but merely that it is not provable *in T*. The reasoning of the last paragraph is correct, but when it is expressed in the language of PA, it cannot be justified from the axioms of T. We will return to this issue later and pinpoint precisely which step in this reasoning cannot be justified in T.

For example, the formal system PA is decidable and, since all of its theorems are true, it is also ω-consistent. Therefore its Gödel sentence, G_{PA}, is true but not provable in PA, so PA is not strong enough to allow us to prove all true statements about the natural numbers. A natural response to this situation is to try to extend PA by adding new axioms, and a natural choice for a new axiom would be the sentence G_{PA}, which we have just seen is true but not provable in PA. So let T_1 be the extension of PA in which G_{PA} is added as a new nonlogical axiom. Then $T_1 \vdash G_{PA}$, but T_1 has its own Gödel sentence G_{T_1}, and by the First Incompleteness Theorem G_{T_1} is true but not provable in T_1. Of course, we could add G_{T_1} as another nonlogical axiom to get a further extension T_2, but the sentence G_{T_2} will be true but unprovable in T_2. As we observed in the previous chapter, the First Incompleteness Theorem implies that no matter how long we continue this process, we will never reach a formal system in which all true statements about the natural numbers are theorems. In fact, we can even define the formal system T_ω to be the extension of PA that has as additional nonlogical axioms the sentence G_{PA} and all of the sentences G_{T_n}, for every n, and the sentence G_{T_ω} will be true but not provable in T_ω.

Another interesting example is the theory T that is the extension of PA in which $\neg G_{PA}$ is added as a new axiom. T is a decidable extension of PA, and the fact that PA $\nvdash G_{PA}$ implies that T is consistent. What is interesting about T is that $T \vdash \neg G_{PA}$, but we have already shown that $\neg G_{PA}$ is false. Thus, this example illustrates that it is possible for a consistent, decidable extension of PA to have false theorems. Once again, this is similar to an observation we made in the previous chapter. (For more on this theory, see exercise 22.)

Another way to think about the meaning of the First Incompleteness Theorem is in terms of nonstandard models of extensions of PA. A *model* of an extension T of PA is an assignment of meanings to the symbols of the language of PA that makes all of the axioms of T come out true. Let us spell out a bit more precisely what this means. First of all, a model $\mathfrak{M}$ must specify a set $U^{\mathfrak{M}}$, called the *universe* of the model. This set determines the meanings that are assigned to the quantifier symbols in the language of PA; the symbols "$\forall$" and "$\exists$" are interpreted as meaning "for every element of $U^{\mathfrak{M}}$" and "for some element of $U^{\mathfrak{M}}$." A model $\mathfrak{M}$ must also specify an element $0^{\mathfrak{M}} \in U^{\mathfrak{M}}$, which determines the meaning of the symbol "0"; the symbol "0" is interpreted as standing for the element $0^{\mathfrak{M}}$.

Next, the model $\mathfrak{M}$ must specify an operation $+^{\mathfrak{M}}$ that can be used to combine any two elements of $U^{\mathfrak{M}}$ to get another element of $U^{\mathfrak{M}}$. This operation is used as the meaning of the symbol "+". Similarly, the model must specify meanings for the symbols "S", "$\cdot$", and "$<$".

Once these meanings have been specified, every sentence in the language of PA will have an interpretation in the model $\mathfrak{M}$. We will say that a sentence is *true in* $\mathfrak{M}$ if it comes out true when its symbols are interpreted as specified by the model $\mathfrak{M}$, and *false in* $\mathfrak{M}$ if it comes out false under this interpretation. (A more detailed explanation of what this means can be found in any logic textbook, but for our purposes this somewhat informal explanation should suffice.) For $\mathfrak{M}$ to be a model of an extension T of PA, all axioms of T must be true in $\mathfrak{M}$.

For example, we can define one model $\mathfrak{M}$ by letting $U^{\mathfrak{M}} = \{0, 1, 2, 3, \ldots\}$, the set of natural numbers, and $0^{\mathfrak{M}} = 0$, and letting $S^{\mathfrak{M}}$, $+^{\mathfrak{M}}$, $\cdot^{\mathfrak{M}}$, and $<^{\mathfrak{M}}$ be the successor operation, the operations of addition and multiplication of natural numbers, and the less-than relation for natural numbers. Of course, this is the model in which all of the symbols are given their intended interpretations. Clearly, all of the axioms of PA are true when the symbols are given their intended interpretations, so $\mathfrak{M}$ is a model of PA. It is sometimes called the *standard model* of PA. According to the Soundness Theorem of first-order logic, all theorems of any formal theory are true in every model of the theory, so all theorems of PA are also true in the standard model. At some points in this chapter we have spoken informally about sentences being true or false. We can now make this talk more precise by saying that we were interpreting the sentences in the standard model of PA.

However, it is a consequence of the First Incompleteness Theorem that there must be other models of PA. To see why, we will use the Completeness Theorem of first-order logic. This theorem was also proven by Gödel (in 1930) and, although the names of the theorems might suggest otherwise, it does not conflict in any way with the Incompleteness Theorems. According to the Completeness Theorem, if T is an extension of PA, P is a sentence in the language of PA, and P is true in every model of T, then $T \vdash P$; equivalently, if $T \nvdash P$, then there must be a model of T in which P is false. For example, suppose that T is a decidable, consistent extension of PA. Then by the First Incompleteness Theorem, $T \nvdash G_T$, so by the Completeness Theorem, there must be a model $\mathfrak{M}$ of T in which G_T is false. But we have already observed that G_T is true – that is, true in the standard model of PA – so $\mathfrak{M}$ must be a nonstandard model.[7]

---

[7]   The reader may be puzzled by the fact that the nonstandard model $\mathfrak{M}$ is not isomorphic to the standard model, for in Chapter 2 we said that the Peano

We might describe the situation as follows. One way to try to characterize the natural numbers is to write down a collection of axioms that are true in the standard model of PA. But if the result is a decidable, consistent extension T of PA, then the attempt at characterizing the natural numbers must fail, because all of the axioms of the theory T will be true in some nonstandard model $\mathfrak{M}$. Furthermore, the model $\mathfrak{M}$ will disagree with the standard model about the truth values of some sentences, such as the sentence G_T. By the Soundness Theorem, all of the theorems of T must be sentences that are true in both the standard model of PA and also the model $\mathfrak{M}$. It follows that if P is any sentence that is true in the standard model but false in $\mathfrak{M}$ (or vice versa), then neither P nor $\neg P$ can be a theorem of T. We might say that the incompleteness of T is caused by the fact that, although T was intended to describe the standard model of PA, it has turned out that it also accurately describes the nonstandard model $\mathfrak{M}$, and these two models disagree about the truth values of some sentences in the language of PA.

The First Incompleteness Theorem applies only to extensions of PA that are decidable, consistent, and, for the second part of the theorem, ω-consistent. Can we escape the limitations imposed by the First Incompleteness Theorem by using an extension of PA that does not have these properties? A few years after Gödel proved the First Incompleteness Theorem, the American logician J. Barkley Rosser (1907–1989) showed

Postulates are categorical; in other words, that all models of the Peano Postulates are isomorphic. The explanation for this apparent contradiction is that the Peano Postulates of Chapter 2 were stated in second-order logic; in particular, the induction axiom (22)(vi) of Chapter 2 involves quantification over all concepts. In this chapter, however, we have worked with axioms stated in first-order logic, so that we could take advantage of the formal systems of logical axioms and rules of inference for first-order logic found in logic textbooks. The first-order version of the induction axiom in Definition 7.4(viii) is weaker than the second-order induction axiom of Chapter 2, because the second-order version applies to *all* concepts, whereas the first-order version applies only to concepts definable by a formula in the first-order language of number theory. The second-order Peano Postulates of Chapter 2 are categorical, but the first-order theory PA of this chapter is not.

The reader may wonder why, then, we have not chosen to use second-order logic in this chapter. The answer is that there is no formal system of axioms and rules of inference in which one can derive just the logical consequences of a collection of axioms stated in the language of second-order logic. Indeed, the First Incompleteness Theorem can be used to show that there could not be such a formal system. For an even stronger result, see Theorem 41C on p. 286 of Enderton (2001).

that if Gödel's sentence G_T is replaced with a somewhat more complicated sentence, then the assumption of ω-consistency in part (2) of the theorem can be changed to ordinary consistency (see exercise 18). Thus, Rosser's proof shows that any decidable, consistent extension of PA is incomplete. Of course, we would not want to use an inconsistent extension of PA; all sentences, including false ones, would be theorems in such a formal system. But why not use an undecidable extension of PA?

To answer this question, we must think about the role played by proofs in the work of mathematicians. When a mathematician claims to have discovered that some mathematical statement P is true, his skeptical colleagues will ask him to justify his claim with a proof. Let us assume that this proof is presented in some formal system T. The purpose of the proof is to convince these colleagues that P is true, but the proof will not accomplish this purpose if the colleagues cannot check the correctness of the proof. Part of checking the correctness of the proof is checking that all sentences that are alleged in the proof to be axioms really are axioms of T. But if T is not decidable then there will be no algorithm that can be used to check whether or not a sentence is an axiom. The result is that a disagreement about the truth of P may not be settled by the proof, but may instead turn into a disagreement about whether or not some statement is an axiom of T. To prevent such disagreements, logicians agree that mathematicians must use formal systems that are decidable.

An example may help to clarify the issue. Consider the extension T of PA that has, as its nonlogical axioms, all true sentences in the language of PA. Every true sentence is a theorem in this formal system. In fact, if P is a true sentence in the language of PA, then there is a one-line proof of P in T. We can simply assert P, justifying our assertion by the fact that P is an axiom of T. But, of course, this proof would not convince anyone who did not already believe that P was true! Thus this formal system would be useless to mathematicians in their everyday work. It is not hard to show that T is ω-consistent and complete, so, by the First Incompleteness Theorem, it must not be decidable. In other words, we have the following interesting consequence of the First Incompleteness Theorem:

COROLLARY 7.19.   *There is no algorithm that determines, for any given sentence of the language of* PA, *whether or not that sentence is true.*

It follows, by the comment after Theorem 7.10, that the set of Gödel numbers of true sentences in the language of PA is not representable. But in fact we can say more: the set of Gödel numbers of true sentences in the language of PA is not even definable in the language of PA.

THEOREM 7.20 (Tarski's Theorem on the Undefinability of Truth).[8] *The set {$n \in \mathbb{N} : n$ is the Gödel number of a true sentence} is not definable in the language of* PA. *In other words, there is no formula $P(x_1)$ such that, for every natural number n, if n is the Gödel number of a true sentence then $P(S^n 0)$ is true, and if not then $P(S^n 0)$ is false.*

Proof. Suppose that $P(x_1)$ were such a formula. By the Fixed Point Lemma, there is a sentence Q such that:

(20) $\quad \text{PA} \vdash Q \leftrightarrow \neg P(\ulcorner Q \urcorner)$.

Thus $Q \leftrightarrow \neg P(S^n 0)$ is true, where $n = \#Q$. But by assumption, if Q is true then $P(S^n 0)$ is true, and if not then $P(S^n 0)$ is false, so $Q \leftrightarrow P(S^n 0)$ is also true. Since these two biconditionals cannot both be true, we have a contradiction. ∎

As another consequence of the First Incompleteness Theorem, we can now also resolve our earlier question about the representability of the set of Gödel numbers of theorems. Let T be any consistent, decidable extension of PA. Then, by part (i) of the First Incompleteness Theorem, G_T is not a theorem of T. It follows that, if the formula $\text{Theorem}_T(x_1)$ represents the set of Gödel numbers of theorems of T, then PA $\vdash \neg\text{Theorem}_T(\ulcorner G_T \urcorner)$. But then, by (17) and the closure of PA under logical consequence, PA $\vdash G_T$, and hence $T \vdash G_T$, contradicting the fact that G_T is not a theorem of T. Therefore, $\text{Theorem}_T(x_1)$ does not represent the set of Gödel numbers of theorems of T. In fact, no formula represents this set:

THEOREM 7.21. *Suppose that T is a consistent extension of* PA. *Then {$n \in \mathbb{N} : n$ is the Gödel number of a theorem of T} is not representable. It follows, by Theorem 7.10, that there is no algorithm that, given a sentence in the language of* PA, *will determine whether or not that sentence is a theorem of T.*

Proof. Suppose that $P(x_1)$ represents the set of Gödel numbers of theorems of T. By the Fixed Point Lemma, there is a sentence Q such that:

---

[8] The theorem is commonly attributed to the Polish logician Alfred Tarski (1902–1983), who was the first to publish it, in 1933. However, as Tarski himself emphasized (pp. 247–8, 277–8), the theorem is foreshadowed in Gödel's 1931 paper, and indeed, Gödel stated and proved the theorem explicitly in a letter to Zermelo in 1931; see *Historia Mathematica*, 6 (1979), 294–304.

(21)   PA $\vdash Q \leftrightarrow \neg P(\ulcorner Q \urcorner)$.

Suppose that $T \vdash Q$. Then, since $P(x_1)$ represents the set of Gödel numbers of theorems of T, PA $\vdash P(\ulcorner Q \urcorner)$. But then, by (21) and the closure of PA under logical consequence, PA $\vdash \neg Q$, so $T \vdash \neg Q$, contradicting the consistency of T. Therefore $T \not\vdash Q$. Applying once again the fact that $P(x_1)$ represents the set of Gödel numbers of theorems of T, we can conclude that PA $\vdash \neg P(\ulcorner Q \urcorner)$. It follows, by (21), that PA $\vdash Q$, so $T \vdash Q$, which is a contradiction.   ∎

As interesting as these consequences of the First Incompleteness Theorem are, they do not directly address Hilbert's program. For this, we need to move on to Gödel's Second Incompleteness Theorem. We begin by picking up where we left off in our project of translating the definitions of this chapter into the language of PA. The last definition we translated was Definition 7.5, so let us consider Definition 7.6.

Suppose that T is a decidable extension of PA. According to Definition 7.6, to say that T is consistent means that there does not exist a sentence P such that both P and $\neg P$ are theorems of T. To translate this definition into the language of PA, we need to be able to talk in that language about one sentence being the negation of another. This is taken care of by our next lemma.

LEMMA 7.22.   *The set* $\{\langle n, m \rangle \in \mathbb{N}^2 : n$ *is the Gödel number of a sentence P and m is the Gödel number of* $\neg P\}$ *is representable.*

Lemma 7.22 can be proven by showing that the set in question is decidable, and then applying Theorem 7.10. What the lemma tells us is that there is a formula $\text{Neg}(x_1, x_2)$ such that if $n = \#P$ and $m = \#\neg P$ for some sentence P, then PA $\vdash \text{Neg}(S^n 0, S^m 0)$, and if not, then PA $\vdash \neg \text{Neg}(S^n 0, S^m 0)$. We can think of $\text{Neg}(x_1, x_2)$ as saying that x_1 is the Gödel number of a sentence and x_2 is the Gödel number of the negation of that sentence.

Using the formula $\text{Neg}(x_1, x_2)$ we can now translate the definition of consistency into the language of PA. Let Con_T be the following sentence:[9]

---

[9]   Of course, by "Theorem$_T(x_2)$" we mean the result of substituting x_2 for all free occurrences of x_1 in Theorem$_T(x_1)$. As was mentioned in footnote 3, this substitution can have unintended consequences if x_2 is not substitutable for x_1 in Theorem$_T(x_1)$. However, this problem can be avoided by renaming bound variables in Theorem$_T(x_1)$ if necessary, and we assume here that this renaming has been done.

(22) $\neg\exists x_1 \exists x_2(\text{Theorem}_T(x_1) \land \text{Theorem}_T(x_2) \land \text{Neg}(x_1, x_2))$.

Then Con_T says that there do not exist two numbers such that both of them are Gödel numbers of theorems of T and one of these theorems is the negation of the other. In other words, Con_T expresses in the language of PA the meaning of the statement "T is consistent."

The key to the Second Incompleteness Theorem is the following relationship between Con_T and the Gödel sentence for T:

LEMMA 7.23.   *If T is a decidable extension of* PA, *then* PA $\vdash \text{Con}_T \rightarrow G_T$.

In fact, it turns out that if T is a decidable extension of PA then PA $\vdash \text{Con}_T \leftrightarrow G_T$, but we will not need this stronger fact. We will put off a careful proof of Lemma 7.23 until later, but we can give a brief explanation of why the lemma is plausible. Earlier, we gave an informal argument showing that if a decidable extension T of PA is consistent, then G_T is true (although unprovable). This informal argument can be translated into the language of PA and all steps of the argument can be justified from the axioms of PA. The result is a formal proof in PA of the sentence $\text{Con}_T \rightarrow G_T$.

Gödel's Second Incompleteness Theorem follows almost immediately from Lemma 7.23, together with the First Incompleteness Theorem:

THEOREM 7.24 (Second Incompleteness Theorem).   *If T is a decidable and consistent extension of* PA, *then* $T \nvdash \text{Con}_T$.

Proof. Suppose that T is a decidable, consistent extension of PA and $T \vdash \text{Con}_T$. By Lemma 7.23, $T \vdash \text{Con}_T \rightarrow G_T$, so by the closure of T under logical consequence, $T \vdash G_T$. But this contradicts part (i) of the First Incompleteness Theorem. Therefore, $T \nvdash \text{Con}_T$.   ∎

Although we have concentrated so far on extensions of PA, the Incompleteness Theorems apply to a wide variety of formal systems. There are many formal systems that are not extensions of PA as we have defined that term, but that can be thought of as strengthenings of PA in some sense. For example, consider the formal system ZFC. The language of ZFC is not the same as the language of PA, and the axioms of PA are not included among the axioms of ZFC, so ZFC is not an extension of PA. However, as we saw in Chapter 3, the basic arithmetical operations on natural numbers can be defined in the language of ZFC, and the Peano Postulates can be proven in ZFC. It follows that all of the theorems of PA can be expressed in the language of ZFC and are provable in ZFC, and it is therefore reasonable to think of ZFC as a strengthening of PA. As a result, we can use reasoning that is similar to the reasoning we have used in our proofs in this chapter to show that the Incompleteness Theorems hold for ZFC. In particular, there is a sentence Con_{ZFC} in the language of ZFC that expresses the statement "ZFC is consistent," and we have the following theorem about this sentence:

THEOREM 7.25 (Second Incompleteness Theorem for ZFC). *If ZFC is consistent, then* ZFC $\nvdash$ Con_{ZFC}.

As we saw in Chapter 3, virtually all of mathematics can be formalized in ZFC, so ZFC is a reasonable choice to play the role of Hilbert's formalized infinitary mathematics. Thus, one way to realize Hilbert's program would be to find a finitary proof of the consistency of ZFC. But such a proof, like other mathematical reasoning, would be formalizable in ZFC, and the formalized version of the proof would be a proof of the sentence Con_{ZFC} in ZFC. According to Theorem 7.25, such a proof cannot exist if ZFC is consistent. Thus, there can be no finitary proof of the consistency of ZFC. It seems likely that the same reasoning would apply to any other formal theory that could serve as a formalization of infinitary mathematics, so, as we observed in the previous chapter, the Second Incompleteness Theorem casts serious doubt on the feasibility of Hilbert's program.

We still owe the reader proofs for two lemmas, 7.15 and 7.23. We will start with the second. It is convenient to begin by identifying the properties of the formula $\text{Theorem}_T(x_1)$ that will be used in the proof. The importance of the three properties in our next lemma was first recognized by Hilbert and the Swiss mathematician Paul Bernays (1888–1977), so they are known as the *Hilbert–Bernays provability properties*. For the second and third of these properties, it is important to specify the formula $\text{Theorem}_T(x_1)$ carefully. In particular, it is not sufficient to assume simply that $\text{Theorem}_T(x_1)$ is the formula $\exists x_2 \text{Proof}_T(x_1, x_2)$, where $\text{Proof}_T(x_1, x_2)$ is some formula that represents the set of pairs of numbers defined in Theorem 7.12. Rather, it is important that $\text{Proof}_T(x_1, x_2)$ be defined as described in (15). (For more on this, see exercise 17.)

LEMMA 7.26. *Suppose that T is a decidable extension of* PA. *Then, for all sentences P and Q:*
  (i)   *If* $T \vdash P$, *then* PA $\vdash$ $\text{Theorem}_T(\ulcorner P \urcorner)$.
  (ii) PA $\vdash$ $\text{Theorem}_T(\ulcorner P \urcorner) \rightarrow \text{Theorem}_T(\ulcorner \text{Theorem}_T(\ulcorner P \urcorner) \urcorner)$.

(iii)   PA ⊢ (Theorem$_T$($\ulcorner P \urcorner$) ∧ Theorem$_T$($\ulcorner P \to Q \urcorner$)) →
Theorem$_T$($\ulcorner Q \urcorner$).

Idea of Proof. Property (i) is simply a restatement of Corollary 7.14, whose proof has already been discussed. Note that since T is an extension of PA, every proof in PA is also a proof in T. Therefore it follows from (i) that if $T \vdash P$ then $T \vdash$ Theorem$_T$($\ulcorner P \urcorner$). This last statement can itself be expressed in the language of PA by the sentence Theorem$_T$($\ulcorner P \urcorner$) → Theorem$_T$($\ulcorner$Theorem$_T$($\ulcorner P \urcorner$)$\urcorner$), and property (ii) says that this sentence is provable in PA. Thus, property (ii) can be proven by showing that the proof of property (i) can itself be formalized in PA. The details of this are long and tedious, and will not be discussed here.[10]

To prove property (iii), note that if we have found proofs of both P and $P \to Q$ in T, then we can construct a proof of Q in T by simply stringing these two proofs together and then adding one additional step in which Q is inferred from P and $P \to Q$. This shows that if $T \vdash P$ and $T \vdash P \to Q$, then $T \vdash Q$. Formalizing this reasoning in PA leads to a proof of property (iii).   ∎

In the proof of Lemma 7.23 it will also be convenient to use the following consequence of the Hilbert–Bernays provability properties:

LEMMA 7.27. *For any sentences P and Q, if* PA $\vdash P \to Q$ *then* PA $\vdash$ Theorem$_T$($\ulcorner P \urcorner$) → Theorem$_T$($\ulcorner Q \urcorner$).

Proof. Suppose PA $\vdash P \to Q$. Then by 7.26(i),

(23)   PA ⊢ Theorem$_T$($\ulcorner P \to Q \urcorner$).

Also, by 7.26(iii),

(24)   PA ⊢ (Theorem$_T$($\ulcorner P \urcorner$) ∧ Theorem$_T$($\ulcorner P \to Q \urcorner$)) →
Theorem$_T$($\ulcorner Q \urcorner$).

The sentence Theorem$_T$($\ulcorner P \urcorner$) → Theorem$_T$($\ulcorner Q \urcorner$) is a logical consequence of the sentences in (23) and (24). Thus, by the closure of PA under logical consequence, PA $\vdash$ Theorem$_T$($\ulcorner P \urcorner$) → Theorem$_T$($\ulcorner Q \urcorner$).   ∎

---

[10]   See Boolos (1979).

We are now ready to give the proof of Lemma 7.23. Many of the steps in this proof are applications of the closure of PA under logical consequence.

Proof of Lemma 7.23. We begin by restating the defining property of G_T:

(25)   PA $\vdash G_T \leftrightarrow \neg\text{Theorem}_T(\ulcorner G_T \urcorner)$.

As a consequence of the left-to-right direction of this biconditional, we have

(26)   PA $\vdash \text{Theorem}_T(\ulcorner G_T \urcorner) \rightarrow \neg G_T$,

and from the right-to-left direction we can conclude that

(27)   PA $\vdash \neg G_T \rightarrow \text{Theorem}_T(\ulcorner G_T \urcorner)$.

By 7.26(ii), we have

(28)   PA $\vdash \text{Theorem}_T(\ulcorner G_T \urcorner) \rightarrow \text{Theorem}_T(\ulcorner \text{Theorem}_T(\ulcorner G_T \urcorner) \urcorner)$.

Applying Lemma 7.27 to (26) gives us

(29)   PA $\vdash \text{Theorem}_T(\ulcorner \text{Theorem}_T(\ulcorner G_T \urcorner) \urcorner) \rightarrow \text{Theorem}_T(\ulcorner \neg G_T \urcorner)$.

We can now combine (27)–(29) to conclude that

(30)   PA $\vdash \neg G_T \rightarrow \text{Theorem}_T(\ulcorner \neg G_T \urcorner)$.

Since the formula $\text{Neg}(x_1, x_2)$ represents the set in Lemma 7.22, we have

(31)   PA $\vdash \text{Neg}(\ulcorner G_T \urcorner, \ulcorner \neg G_T \urcorner)$.

Combining (27), (30), and (31), we obtain

(32)   PA $\vdash \neg G_T \rightarrow (\text{Theorem}_T(\ulcorner G_T \urcorner) \wedge \text{Theorem}_T(\ulcorner \neg G_T \urcorner) \wedge \text{Neg}(\ulcorner G_T \urcorner, \ulcorner \neg G_T \urcorner))$.

Thus,

(33)   PA $\vdash \neg G_T \rightarrow \exists x_1 \exists x_2 (\text{Theorem}_T(x_1) \wedge \text{Theorem}_T(x_2) \wedge \text{Neg}(x_1, x_2))$.

But, according to (22), Con$_T$ is simply the negation of the right-hand side of the conditional in (33). Thus,

(34) PA $\vdash \neg G_T \rightarrow \neg$Con$_T$.

Finally, by contraposition we can conclude that PA $\vdash$ Con$_T \rightarrow G_T$, as required. ∎

Now we turn to the proof of Lemma 7.15, the Fixed Point Lemma. The Fixed Point Lemma says that a sentence in the language of PA can, in a sense, talk about itself. To prepare for the proof it may be helpful to consider an example, due to Quine, illustrating how a sentence in English can talk about itself.[11] We will work up to Quine's example gradually.

Consider the following two sentences:

(35) philosophy of mathematics is a fascinating subject'.

(36) "philosophy of mathematics" is three words long.

The subject matter of sentence (35) is the philosophy of mathematics, and the sentence makes the true assertion that it is a fascinating subject. The quotation marks in sentence (36) indicate that its subject matter is not the philosophy of mathematics but, rather, the phrase "philosophy of mathematics", and the sentence makes the true assertion that this phrase consists of three words.

Here is another example of a sentence whose subject is a phrase:

(37) "is three words long" is three words long.

Of course, this sentence is false, since the phrase "is three words long" is actually four words long. Sentence (37) has the peculiar form of a phrase appended to its own quotation, so we can express the fact that (37) is false as follows:

(38) "is three words long" yields a falsehood when appended to its own quotation.

Sentence (38) says that when the phrase "is three words long" is appended to its own quotation, the result is a falsehood. But the result

---

[11] This example can be found in W. V. Quine's "Paradox" (1962; in Quine, 1976); see p. 17.

of appending the phrase "is three words long" to its own quotation is sentence (37). Thus, sentence (38) talks in a roundabout way about sentence (37), and what it says about sentence (37) is that it is false. Since sentence (37) is, indeed, false, sentence (38) is true.

Here, finally, is Quine's example:

(39)   "yields a falsehood when appended to its own quotation" yields a falsehood when appended to its own quotation.

Like sentence (38), sentence (39) talks about the result of appending a certain phrase to its own quotation. But in this case, the result is sentence (39) itself! Thus, sentence (39) talks in a roundabout way about itself, and what it says about itself is that it is false. We will leave it to the reader to decide whether sentence (39) is true or false. Notice that by replacing the word "falsehood" in sentence (39) by another word or phrase we could construct a sentence saying whatever we wanted about itself. The proof of the Fixed Point Lemma will involve transferring the ideas behind sentence (39) to the language of PA.

For technical reasons that will become clear later, it will be convenient to work with formulas in the language of PA whose only free variable is x_2. Suppose $R(x_2)$ is such a formula, and let n be its Gödel number. If we substitute the numeral $S^n 0$ for all free occurrences of x_2 in $R(x_2)$, then we get the sentence $R(S^n 0)$. We will call this sentence the *diagonalization* of the formula $R(x_2)$. The operation of forming the diagonalization of a formula will play a role in our proof of the Fixed Point Lemma that corresponds to the role played in Quine's example by the operation of appending a phrase to its own quotation.

For example, suppose $R(x_2)$ is the formula "$(x_2 = 0)$". Its Gödel number is

$$(40) \quad \#R(x_2) = \#\langle 7,\ 16,\ 9,\ 10,\ 8 \rangle = 2^8 \cdot 3^{17} \cdot 5^{10} \cdot 7^{11} \cdot 11^9$$
$$= 1,505,268,403,034,293,837,167,718,297,500,$$
$$000,000.$$

Thus, the diagonalization of $R(x_2)$ is the formula "$(SSS \ldots 0 = 0)$", where the "$\ldots$" stands for 1,505,268,403,034,293,837,167,718,297,499,999, 997 more occurrences of "S".

To imitate the construction of Quine's sentence, which talks about the operation of appending a phrase to its own quotation, we will need to be able to talk in the language of PA about the diagonalization of formulas. Our next lemma ensures that this is possible.

LEMMA 7.28. *The set* $\{\langle m, n \rangle \in \mathbb{N}^2 :$ *n is the Gödel number of a formula whose only free variable is x_2, and m is the Gödel number of the diagonalization of that formula}* is representable.

Proof. Our definition of the diagonalization operation gives an algorithm for finding the diagonalization of any given formula whose only free variable is x_2. It follows that the set in the lemma is decidable, and therefore by Lemma 7.10 it is also representable. ∎

Lemma 7.28 tells us that there is a formula $\text{Diag}(x_1, x_2)$ such that if n is the Gödel number of a formula $R(x_2)$ and m is the Gödel number of the sentence $R(S^n 0)$, then PA $\vdash \text{Diag}(S^m 0, S^n 0)$, and if not then PA $\vdash \neg\text{Diag}(S^m 0, S^n 0)$. Intuitively, we can think of $\text{Diag}(x_1, x_2)$ as saying that x_2 is the Gödel number of a formula, and x_1 is the Gödel number of the diagonalization of that formula. We will find it useful to work not with the formula $\text{Diag}(x_1, x_2)$ but, rather, with the following related formula, which we will call $D(x_1, x_2)$:

(41) $\text{Diag}(x_1, x_2) \wedge \neg\exists x_3(x_3 < x_1 \wedge \text{Diag}(x_3, x_2))$.

Intuitively, $D(x_1, x_2)$ says that x_2 is the Gödel number of a formula, and x_1 is the *smallest* number that is the Gödel number of the diagonalization of that formula. The reason for using $D(x_1, x_2)$ rather than $\text{Diag}(x_1, x_2)$ is given by the following lemma:

LEMMA 7.29. *Suppose n is the Gödel number of a formula $R(x_2)$ and m is the Gödel number of its diagonalization, $R(S^n 0)$. Then*:

PA $\vdash \forall x_1(D(x_1, S^n 0) \leftrightarrow x_1 = S^m 0)$.

Proof. By Lemma 7.28, we already know that

(42) PA $\vdash \text{Diag}(S^m 0, S^n 0)$

and

(43) for each natural number $k < m$, PA $\vdash \neg\text{Diag}(S^k 0, S^n 0)$.

Notice that (43) represents not just one theorem of PA, but a list of m such theorems, one for each value of k from 0 to $m - 1$. It can also be shown that

(44) $PA \vdash \forall x_3(x_3 < S^m0 \rightarrow (x_3 = 0 \lor x_3 = S0 \lor x_3 = SS0 \lor \ldots$
$\lor x_3 = S^{m-1}0))$.

(For an example of this, see exercise 6(b).) By the closure of PA under logical consequence, it follows from (43) and (44) that

(45) $PA \vdash \neg \exists x_3(x_3 < S^m0 \land \text{Diag}(x_3, S^n0))$.

Combining this with (42) gives us

(46) $PA \vdash \text{Diag}(S^m0, S^n0) \land \neg \exists x_3(x_3 < S^m0 \land \text{Diag}(x_3, S^n0))$.

But according to (41), the sentence in (46) is just $D(S^m0, S^n0)$, so we have

(47) $PA \vdash D(S^m0, S^n0)$,

or, equivalently,

(48) $PA \vdash \forall x_1(x_1 = S^m0 \rightarrow D(x_1, S^n0))$.

This is the right-to-left direction of the biconditional in the lemma.

   For the left-to-right direction, we will once again use (45). We begin by rephrasing it in the following equivalent form:

(49) $PA \vdash \forall x_1(x_1 < S^m0 \rightarrow \neg \text{Diag}(x_1, S^n0))$.

We also have, as a consequence of (42), that

(50) $PA \vdash \forall x_1(S^m0 < x_1 \rightarrow \exists x_3(x_3 < x_1 \land \text{Diag}(x_3, S^n0)))$.

According to (41), the right-hand sides of the implications in both (49) and (50) contradict $D(x_1, S^n0)$. Therefore,

(51) $PA \vdash \forall x_1(x_1 < S^m0 \rightarrow \neg D(x_1, S^n0))$

and

(52) $PA \vdash \forall x_1(S^m0 < x_1 \rightarrow \neg D(x_1, S^n0))$.

Finally, we will use the fact that the trichotomy law can be proven in PA (see exercise 7), from which it follows that

(53) $PA \vdash \forall x_1(x_1 < S^m0 \lor S^m0 < x_1 \lor x_1 = S^m0)$.

We can now apply the closure of PA under logical consequence to (51)–(53) to conclude that

(54) PA $\vdash \forall x_1(D(x_1, S^n0) \rightarrow x_1 = S^m0)$,

which is the left-to-right direction of the required biconditional. ∎

We now have everything we need to prove the Fixed Point Lemma.

Proof of Lemma 7.15. Recall that we are given a formula $P(x_1)$, and we must find a sentence Q such that PA $\vdash Q \leftrightarrow P(\ulcorner Q \urcorner)$. Let us say that a sentence Q is a *P-sentence* if the sentence $P(\ulcorner Q \urcorner)$ is true. Then we might say informally that we are looking for a sentence Q that says "I am a *P*-sentence."

We begin by defining a formula $R(x_2)$ that will correspond to Quine's phrase "yields a falsehood when appended to its own quotation". Informally, $R(x_2)$ will say "x_2 is the Gödel number of a formula that yields a *P*-sentence when it is diagonalized." Formally, we define $R(x_2)$ to be the following formula:

(55) $\exists x_1(D(x_1, x_2) \wedge P(x_1))$.

Quine's example is the result of appending his phrase to its own quotation. By analogy, to define the sentence Q we form the diagonalization of $R(x_2)$. Let n be the Gödel number of $R(x_2)$, and let Q be the sentence $R(S^n0)$. Thus, Q is the sentence

(56) $\exists x_1(D(x_1, S^n0) \wedge P(x_1))$.

Let m be the Gödel number of Q. Then, according to Lemma 7.29,

(57) PA $\vdash \forall x_1(D(x_1, S^n0) \leftrightarrow x_1 = S^m0)$.

Thus,

(58) PA $\vdash \exists x_1(D(x_1, S^n0) \wedge P(x_1)) \leftrightarrow \exists x_1(x_1 = S^m0 \wedge P(x_1))$.

But the left side of the biconditional in (58) is just Q, so we have

(59) PA $\vdash Q \leftrightarrow \exists x_1(x_1 = S^m0 \wedge P(x_1))$,

or, more simply,

(60) $\text{PA} \vdash Q \leftrightarrow P(S^m 0)$.

Since m is the Gödel number of Q, this means that $\text{PA} \vdash Q \leftrightarrow P(\ulcorner Q \urcorner)$, as required.   ■

Exercises

1. Write formulas in the language of PA that, in the intended interpretation for the language, have the following meanings. Be sure to include all parentheses required by Definitions 7.1 and 7.2.
    (a)   Addition is commutative.
    (b) x_1 is a perfect square.
    (c) x_1 and x_2 are relatively prime; that is, there is no number larger than 1 that they are both divisible by.
2. Find a proof in PA of the sentence $((S0 \cdot S0) = S0)$.
3. Show that $\text{PA} \vdash (S0 < SS0)$. (You do not need to find a proof; just show that $(S0 < SS0)$ follows logically from the PA axioms, and apply the closure of PA under logical consequence.)
4. Show that $\text{PA} \vdash \neg(SS0 < S0)$. (Hint: You may find it useful to show first that $\text{PA} \vdash \forall x_1 \neg(SS0 + x_1 = 0)$. Use mathematical induction for this. In other words, use an instance of PA axiom (viii).)
5. Show that the associative and commutative laws for addition are theorems of PA. In other words:
    (a)   Show that $\text{PA} \vdash \forall x_1 \forall x_2 \forall x_3((x_1 + x_2) + x_3 = x_1 + (x_2 + x_3))$. (Hint: Imitate the proof of Theorem 3.11.)
    (b)   Show that $\text{PA} \vdash \forall x_1 \forall x_2(x_1 + x_2 = x_2 + x_1)$. (Hint: Imitate the solution to exercise 5 of Chapter 3.)
6. (a)   Show that $\text{PA} \vdash \forall x_1 \neg(x_1 < x_1)$. (Hint: Use mathematical induction.)
    (b)   Show   that $\text{PA} \vdash \forall x_1(x_1 < SS0 \rightarrow (x_1 = 0 \lor x_1 = S0))$. (Hint: Use mathematical induction.)
7. Show   that   the   trichotomy   law   is   a   theorem   of   PA: $\text{PA} \vdash \forall x_1 \forall x_2(x_1 < x_2 \lor x_2 < x_1 \lor x_1 = x_2)$. (Hint: Imitate the solution to exercise 8 of Chapter 3.)
8. Let $E = \{\langle m, n \rangle \in \mathbb{N}^2 : m = n\}$. Show that E is representable.
9. Let $A = \{\langle t, n \rangle \in \mathbb{N}^2 : t$ is a sequence code number, and n is the sum of the numbers in the sequence coded by $t\}$. For example, $\langle \#\langle 2, 1, 3 \rangle, 6 \rangle \in A$, but $\langle \#\langle 2, 3 \rangle, 6 \rangle \notin A$. Use the idea of the proof of Theorem 7.10 to explain how to find a formula that represents A. (You do not need to prove that the formula represents A.)

10. Explain why every representable set is decidable.
11. Compute the Gödel numbers of the following expressions:
    (a) $x_3 +)\neg 0$. (Of course, this is not a formula; it is just a meaningless sequence of symbols. But it still has a Gödel number.)
    (b) $(x_1 \cdot S0) = x_1$.
12. Determine whether the following numbers are Gödel numbers of expressions, and if so what expressions they are the Gödel numbers of:
    (a) 106,288,200,000,000,000.
    (b) 13,720,000,000.
    (c) 161,252,785,673,234,064,776,494,200,382,249,149,347, 524,097,765,963,394,383,821,290,729,762,920,307,556, 493,325,000,000 $= 2^6 \cdot 3^{17} \cdot 5^8 \cdot 7^{17} \cdot 11^{13} \cdot 13^{12} \cdot 17^{17} \cdot 19^9$.
13. By Theorem 7.11, there is a formula $P(x_1)$ that represents the set $\{n \in \mathbb{N} : n$ is the Gödel number of a formula$\}$. By the Fixed Point Lemma, there is a sentence Q such that PA $\vdash Q \leftrightarrow \neg P(\ulcorner Q \urcorner)$. (Thus, we might say informally that Q says "I am not a formula.") Is Q a theorem of PA? Is $\neg Q$ a theorem of PA?
14. Prove that if T is a decidable, ω-consistent extension of PA, then $T \nvdash G_T \rightarrow \text{Theorem}_T(\ulcorner G_T \urcorner)$.
15. Let us suppose that T is a decidable extension of PA. By the Fixed Point Lemma, there is a sentence P such that PA $\vdash P \leftrightarrow \text{Theorem}_T(\ulcorner P \urcorner)$. Thus, P says "I am a theorem of T." Show that $T \vdash P$. (Thus, P is true: it says that it is a theorem of T, and it is! You might want to compare this to exercise 2(c) of Chapter 6.) More generally, prove Löb's Theorem: if P is any sentence such that $T \vdash \text{Theorem}_T(\ulcorner P \urcorner) \rightarrow P$, then $T \vdash P$. (Hint: By the Fixed Point Lemma, there is a sentence Q such that PA $\vdash Q \leftrightarrow (\text{Theorem}_T(\ulcorner Q \urcorner) \rightarrow P)$. Show that the following statements are true:

    PA $\vdash \text{Theorem}_T(\ulcorner Q \urcorner) \rightarrow \text{Theorem}_T(\ulcorner \text{Theorem}_T(\ulcorner Q \urcorner) \rightarrow P \urcorner)$.

    PA $\vdash \text{Theorem}_T(\ulcorner Q \urcorner) \rightarrow \text{Theorem}_T(\ulcorner P \urcorner)$.

    $T \vdash \text{Theorem}_T(\ulcorner Q \urcorner) \rightarrow P$.

    $T \vdash Q$.

    Finally, use the last two statements to conclude that $T \vdash P$.)
16. Suppose T is a decidable extension of PA. For any natural number n, we will say that T is *consistent up to n* if there is no sentence P such that $T \vdash P, T \vdash \neg P$, and the Gödel numbers of $P, \neg P$, and

proofs of P and $\neg P$ are all smaller than n. We can express this in the language of PA by defining $\text{ConUpTo}_T(x_5)$ to be the formula:

$$\neg\exists x_1 \exists x_2 \exists x_3 \exists x_4 (x_1 < x_5 \land x_2 < x_5 \land x_3 < x_5 \land x_4 < x_5 \land$$
$$\text{Proof}_T(x_1, x_2) \land \text{Proof}_T(x_3, x_4) \land \text{Neg}(x_1, x_3)).$$

Then $\text{ConUpTo}_T(x_5)$ says that T is consistent up to x_5. Show that if T is consistent then, for every $n \in \mathbb{N}$, PA $\vdash \text{ConUpTo}_T(S^n 0)$, but $T \nvdash \forall x_5 \text{ConUpTo}_T(x_5)$. (Hint: Use the fact that, for every natural number n,

$$\text{PA} \vdash \forall x_1 (x_1 < S^n 0 \rightarrow (x_1 = 0 \lor x_1 = S0 \lor x_1 = SS0 \lor$$
$$\dots \lor x_1 = S^{n-1} 0)).$$

See (44) in the proof of Lemma 7.29, and exercise 6(b).)

17. Suppose T is a decidable, consistent extension of PA. Let $\text{Proof}'_T(x_1, x_2)$ be the formula:

$$\text{Proof}_T(x_1, x_2) \land \exists x_5 (x_1 < x_5 \land x_2 < x_5 \land \text{ConUpTo}_T(x_5)),$$

where $\text{ConUpTo}_T(x_5)$ is defined as in exercise 16. Let $\text{Theorem}'_T(x_1)$ and Con'_T be defined from $\text{Proof}'_T(x_1, x_2)$ in the same way that $\text{Theorem}_T(x_1)$ and Con_T were defined from $\text{Proof}_T(x_1, x_2)$. In other words, let $\text{Theorem}'_T(x_1)$ be the formula

$$\exists x_2 \text{Proof}'_T(x_1, x_2),$$

and let Con'_T be the sentence

$$\neg\exists x_1 \exists x_2 (\text{Theorem}'_T(x_1) \land \text{Theorem}'_T(x_2) \land \text{Neg}(x_1, x_2)).$$

(a) Show that $\text{Proof}'_T(x_1, x_2)$ represents $\{\langle n, m \rangle \in \mathbb{N}^2 : n \text{ is the Gödel number of a sentence and } m \text{ is the Gödel number of a proof of that sentence in } T\}$.

(b) Show that $T \vdash \text{Con}'_T$. (Hint: According to its definition, Con'_T is logically equivalent to

$$\neg\exists x_1 \exists x_2 \exists x_3 \exists x_4 (\text{Proof}_T(x_1, x_2) \land \text{Proof}_T(x_3, x_4) \land \text{Neg}(x_1, x_3)$$
$$\land \exists x_5 (x_1 < x_5 \land x_2 < x_5 \land \text{ConUpTo}_T(x_5)) \land \exists x_6 (x_3 < x_6$$
$$\land x_4 < x_6 \land \text{ConUpTo}_T(x_6))).$$

Use the trichotomy law (see exercise 7) to show that

$$PA \vdash \neg Con'_T \rightarrow \exists x_1 \exists x_2 \exists x_3 \exists x_4 \exists x_5 (Proof_T(x_1, x_2)$$
$$\wedge\ Proof_T(x_3, x_4) \wedge Neg(x_1, x_3) \wedge x_1 < x_5 \wedge x_2 < x_5$$
$$\wedge\ x_3 < x_5 \wedge x_4 < x_5 \wedge ConUpTo_T(x_5)).$$

Equivalently,

$$PA \vdash \neg Con'_T \rightarrow \exists x_5[ConUpTo_T(x_5) \wedge \exists x_1 \exists x_2 \exists x_3 \exists x_4(x_1 < x_5$$
$$\wedge\ x_2 < x_5 \wedge x_3 < x_5 \wedge x_4 < x_5 \wedge Proof_T(x_1, x_2)$$
$$\wedge\ Proof_T(x_3, x_4) \wedge Neg(x_1, x_3))].$$

Finally, combine this with the definition of $ConUpTo_T(x_5)$ to conclude that $PA \vdash Con'_T$.)

18. Prove Rosser's Theorem: if T is a decidable, consistent extension of PA, then there is a sentence R_T such that $T \nvdash R_T$ and $T \nvdash \neg R_T$. (Hint: Suppose T is a decidable, consistent extension of PA. Define the Rosser sentence R_T in exactly the same way that we defined the Gödel sentence G_T, except using $Theorem'_T(x_1)$ instead of $Theorem_T(x_1)$, where $Theorem'_T(x_1)$ is defined as in exercise 17. In other words, use the Fixed Point Lemma to choose R_T so that

$$PA \vdash R_T \leftrightarrow \neg Theorem'_T(\ulcorner R_T \urcorner).$$

The proof that $T \nvdash R_T$ is similar to the proof that $T \nvdash G_T$. To show that $T \nvdash \neg R_T$, suppose that $T \vdash \neg R_T$ and then prove that $T \vdash \neg Con'_T$. By exercise 17(b), this contradicts the consistency of T.)

19. Suppose T is a decidable, consistent extension of PA. Let us say that a sentence P is *efficiently provable* in T if there is a proof of P in T whose Gödel number is less than $10^{10^{\#P}}$. Let $E = \{n \in \mathbb{N} : n$ is the Gödel number of a sentence that is efficiently provable in $T\}$.

    (a) Show that E is representable. (Hint: Show that E is decidable, and apply Theorem 7.10.)

    (b) Show that there is a sentence Q such that $PA \vdash Q$, but Q is not efficiently provable in T. (Hint: By part (a), let $EfficientlyProvable_T(x_1)$ be a formula that represents E. By the Fixed Point Lemma, let Q be a sentence such that

$$PA \vdash Q \leftrightarrow \neg EfficientlyProvable_T(\ulcorner Q \urcorner).$$

Now assume that Q is efficiently provable in T, and derive a contradiction.)

20. Suppose T is a decidable extension of PA. The Second Incompleteness Theorem says that if T is consistent, then $T \not\vdash \text{Con}_T$. This statement can itself be expressed in the language of PA by the sentence

$$\text{Con}_T \rightarrow \neg\text{Theorem}_T(\ulcorner\text{Con}_T\urcorner).$$

Show that this sentence is a theorem of PA; that is, PA $\vdash$ $\text{Con}_T \rightarrow \neg\text{Theorem}_T(\ulcorner\text{Con}_T\urcorner)$. (Hint: Use Lemma 7.23.)

21. Suppose T is a decidable extension of PA. Show that if T is ω-consistent then $T \not\vdash \text{Con}_T \rightarrow \neg\text{Theorem}_T(\ulcorner\neg\text{Con}_T\urcorner)$. (Hint: Suppose $T \vdash \text{Con}_T \rightarrow \neg\text{Theorem}_T(\ulcorner\neg\text{Con}_T\urcorner)$. Apply Löb's Theorem (exercise 15) to show that $T \vdash \neg\text{Con}_T$. Conclude that $T \vdash \exists x_5 \neg\text{ConUpTo}_T(x_5)$, where $\text{ConUpTo}_T(x_5)$ is defined as in exercise 16. Combine this with exercise 16 to conclude that T is not ω-consistent.)

22. Let T be the theory whose axioms are those of PA, together with the sentence $\neg G_{\text{PA}}$. As was discussed in the text, T is consistent. Show that T is not ω-consistent.

23. Let T be a decidable, consistent extension of PA. We will say that a formula $F(x_1)$ is a *T-name* for a natural number n if $T \vdash \forall x_1(F(x_1) \leftrightarrow x_1 = S^n0)$. For example, the formula $(x_1 = (SSS0 + SS0))$ is a PA-name for 5.[12]

    (a) Show that no formula can be a T-name for two different natural numbers.

    (b) Show that, for every natural number k, there are only finitely many natural numbers that have a T-name that is k symbols long. (Hint: Although there are infinitely many formulas of some lengths, many of them are logically equivalent to each other.)

    (c) Show that $\{\langle n, k\rangle \in \mathbb{N}^2 : n$ has a T-name that is k symbols long$\}$ is definable in the language of PA. In other words, show that there is a formula $\text{Nameable}_T(u, v)$ such that, for all natural numbers n and k, $\text{Nameable}_T(S^n0, S^k0)$ is true (that is, true in the standard model) if and only if n has a T-name that is k symbols long. (Hint: You might want to start with the fact that $\{\langle n, m, p\rangle \in \mathbb{N}^3 : m$ is the Gödel number of a formula

---

[12] This exercise is based on the proof in Boolos (1989). You might want to compare it to exercise 3 of Chapter 6, which is based on Berry's paradox.

$F(x_1)$ and p is the Gödel number of a proof in T of $\forall x_1(F(x_1) \leftrightarrow x_1 = S^n 0)\}$ is decidable, and therefore representable.)

Now let $B(x_1, x_2)$ be the formula:

$$\exists v(v = S^2 0 \cdot x_2 \wedge \neg \text{Nameable}_T(x_1, v) \wedge \forall u(u < x_1 \rightarrow \text{Nameable}_T(u, v))).$$

Thus, $B(x_1, x_2)$ says that x_1 does not have a T-name of length $2 \cdot x_2$, but every number less than x_1 does have such a T-name. In other words, $B(x_1, x_2)$ says that x_1 is the smallest natural number that does not have a T-name of length $2 \cdot x_2$. Let k be the length of the formula $B(x_1, x_2)$, and let $F(x_1)$ be the formula $B(x_1, S^k 0)$; in other words, $F(x_1)$ is the formula

$$\exists v(v = S^2 0 \cdot S^k 0 \wedge \neg \text{Nameable}_T(x_1, v) \wedge \forall u(u < x_1 \rightarrow \text{Nameable}_T(u, v))).$$

(d) Show that there is a number n such that $\forall x_1(F(x_1) \leftrightarrow x_1 = S^n 0)$ is true.

(e) Letting n be the number from part (d), show that $T \nvdash \forall x_1(F(x_1) \leftrightarrow x_1 = S^n 0)$. This establishes that $\forall x_1(F(x_1) \leftrightarrow x_1 = S^n 0)$ is a sentence that is true, but not provable in T.

8

Coda

Our intellectual adventure has not been without frustration. Just as a seductive conception is presented which holds out the prospect of settling some of our original questions about mathematics, we discover that, for one reason or another, its leading idea is ultimately unsatisfactory or unrealizable. We are repeatedly drawn in, only to meet with disappointment. As someone said of Lord Berners' paintings, the delight we take in them is tempered by the regret that being so good they are not just a little better.

Logicism, had it been successfully prosecuted as Frege envisaged, would have relieved a number of the mysteries about mathematics that we discussed in Chapter 1. The *a priori* nature of mathematics, for the logicist, is a consequence of the complete generality of logical knowledge; as mathematics is about nothing in particular, it is little wonder that nothing in particular needs to be observed in order to acquire mathematical knowledge. Logicism also makes it obvious why proof is so central in mathematics, since deduction is the obvious route to logical knowledge. And the universal applicability of mathematics is likewise an easy consequence of the fact that mathematics, being really logic in disguise, is thoroughly general: being about nothing, logic is really about everything, and so the ubiquitous relevance of mathematics is only to be expected. Finally, if logicism had been successful, then our grasp of the infinite could have been seen, not as something alien to our finite natures but, rather, as something that was integral to our rational natures. For, according to logicism, to grasp the laws of thought is, in perhaps a cloaked fashion, to apprehend the infinite.

Unfortunately, the infinities of the mathematical world are the rocks upon which Frege's logicism foundered. For we saw in Chapter 4 that Russell's Paradox reveals a tension between the logicist's goal of doing justice to the completed infinite of the classical mathematician and the logicist's desire to do so by reducing it to a system that is completely general in its applicability. In Chapter 3, we explored how this goal might be reached without satisfying this desire, by showing how the resources of set theory can effect a reduction of analysis. This reduction involves, of course, a substantial philosophical compromise, for set theory's lack of generality leaves it prey to many of the questions that troubled us about mathematics in the first place. And in ZFC the existence of infinite sets was not proven, but was simply accepted as an axiom.

That said, there are respects in which Frege's logicism is a tremendous achievement. For it has provided us with a wealth of philosophical reflections on the nature of language and mathematics that continue to inspire, as well as with formal logic, both a tool and a conceptual analysis of the first order. Frege wished to discover the source of mathematical knowledge and, while he did not find there what he expected, he gave us the resources with which to continue the search. Frege made it possible to ask that central question and to evaluate answers to it, and these alone are significant achievements. As it turns out, study of the very tools that made his project possible leads to the verdict that it was ultimately unrealizable: Gödel's far-reaching work on the nature of logic and formal systems reveals that there can be no single consistent theory from which all mathematical truths can be derived. There is some irony in the fact that while Gödel's great results put paid to Frege's dream, they owe their own existence to it: for it was Frege's work that led to the extraordinary refinement in our understanding of logic and formal systems upon which the Incompleteness Theorems rest.[1]

Of course, the modern set-theoretical reduction, which, while not Frege's, employs many of his mathematical contributions, is not without interest itself, both philosophical and mathematical. While set theory does not exhibit the generality that Frege hankered for, it is still highly general in that a set can have *anything* as elements. Whether there are ways in which a reduction to set theory might dispel any philosophical clouds over mathematics is a matter for continuing debate. In particular, any attempt at using this reduction to resolve philosophical worries

---

[1] As noted earlier, the Peano Postulates can be derived from the equivalence between statements (16) and (17) of Chapter 2. This equivalence is sometimes called *Hume's Principle*, because Frege introduced it with a quotation from Hume (*Foundations*, §63), and the derivability of Peano's Postulates from Hume's Principle in second-order logic has been dubbed *Frege's Theorem*. Although Frege would ultimately have found little of philosophical interest in this theorem (because he did not take Hume's Principle to offer an adequate definition of the natural members), some contemporary philosophers have valued it more highly. See Boolos (1998), Demopoulos (1995), and Hale and Wright (2001).

about mathematics must take account of the incompleteness of ZFC. Gödel's theorem shows that if ZFC is consistent, then it cannot be complete, and in fact, as we saw at the end of Chapter 3, there are interesting mathematical statements, such as the continuum hypothesis, that are known to be neither provable nor refutable in ZFC, assuming that theory to be consistent. Do these statements have truth values? Their truth values are not determined by the set-theoretic reduction; if they have truth values, what determines these values? And if they do not, how can we justify the use in set theory of the Law of the Excluded Middle, which asserts that every statement is either true or false?

This brings us to intuitionism. It too holds out the promise of easing some of our perplexity about mathematics. The intuitionist rejects the idea that mathematical truth holds independently of us, that it is determined by how matters stand in some separately existing realm. Rather, mathematical truth is strictly linked to human activity, in particular to the practice of proving propositions. As such, our capacity to know mathematical truths seems less mysterious than it might otherwise, for it is a matter of knowing about our own doings, about what we ourselves are up to. Furthermore, since intuitionistic semantics is given in terms of proof, not truth, it is obvious why that particular route to knowledge should play the role it does in the justification of mathematical claims. And of course, intuitionism addresses the problem of the infinite by banishing the actual infinite and countenancing only the potential infinite, which it is reasonable to suppose is within our ken.

The issue of applicability is less clear-cut. We saw in Chapter 5 that a devout intuitionist will not accept all of classical mathematics. One naturally suspects that the intuitionist will therefore find it especially difficult to account for the applicability of mathematics, for no doubt much of the mathematics employed in scientific theories is intuitionistically meaningless. The intuitionist appears far worse off in this respect than the classical mathematician, who can at least understand the mathematics whose applicability in natural science can seem so mysterious.

But this is not decisive, for what we need to know is whether the mathematics that finds itself applicable to the natural world is modifiable so as to make it intuitionistically acceptable. These are matters that are not presently well-understood. One reason for thinking that this modification might be possible is the nature of measurement, the means whereby our scientific accounts of the world make contact with reality. A crucial component of our confidence in an empirical theory is its ability to make sense of what we observe, where these observations consist largely of reports of measurements. This is significant because measurement is always, by its nature, finite: while it might be that we have the capacity, in principle, to measure some phenomenon to any degree of accuracy, each actual measurement achieves only a finite degree of accuracy. We have seen that this property, that of being actually finite though capable of extension without limit, characterizes the intuitionistic conception of the infinite. Perhaps this offers some reason to think that the mathematical theories embedded within science can be recast intuitionistically without loss to the predictive or explanatory powers of the scientific accounts. The physical world has no room for the completed infinite, so why think that theories of that world will need to be supported by a mathematical scaffolding that is itself committed to such a conception of infinity?

And what of the uncanny utility of mathematics in science? Focusing on measurement might also help to make the empirical applicability of mathematics more understandable. According to the intuitionist, the basis of mathematics' content is the always actually finite but forever extendable activity of the mathematician. Clearly, this activity has something important in common with the very practice that gives scientific theories *their* content, namely that of taking the measure of the empirical world. Might this fundamental kinship of their respective sources of content shed some light on mathematics' applicability to the natural world? Again, all these matters are as yet unresolved.

In evaluating Hilbert's program, we must as ever distinguish between the promise and the present. Had it been successful, the task of explaining mathematical knowledge would have been much simplified. For knowledge of infinitary mathematics could then have been seen to be simply an illusion. Infinitary mathematics is really meaningless, so there is nothing to know. Of course, the problem of explaining knowledge of finitary mathematics remains. However Hilbert thought that could be done, it is plausibly reckoned a far more tractable problem than that of accounting for knowledge of classical mathematics, or even of intuitionistic mathematics. The role of classical proof would have been much clarified as well: for Hilbert, it was not a method of deriving truths from truths but, rather, a convenient technique for carrying out the manipulation of symbols in a formal system. The consistency proof would have shown that such manipulations can be trusted to lead only to true finitary statements, which explains why mathematicians use them. As for the applicability of mathematics, a Hilbertian finitist might well have insisted that most scientific theories, in light of their dependence on infinitary mathematics, are meaningless. They are not correct descriptions of reality, but instead calculating devices that usefully predict the results of measurement; the only statements of the theory that are meaningful are those that make some finitary claim. The problem of explaining why classical mathematics provides the means to say true things about the world therefore simply does not arise. It is replaced by the constructive task of explaining why the meaningless mathematical apparatus embedded in scientific theories can be relied upon to take one from correct measurement to correct measurement. And presumably, had Hilbert's project been a success, this task would have appeared quite within reach.

Of course, Hilbert's project is unrealizable. One response has simply been to cut loose the idea of justification. That is, one still regards infinitary mathematics as meaningless, and proof as merely a formal symbol-manipulation game that is intended to yield correct predictions of finite calculations. But one now recognizes that there is no guarantee that such manipulations will always work; there is only the fact that they have worked well so far. On this view, mathematics becomes something of a risky game. If one is enough of a platonist to believe that there is a right answer about the consistency of ZFC despite our inability to settle the matter (this requires platonism only for the countable universe of finite sequences of symbols), then one will believe that ZFC really is either consistent or inconsistent. *If* ZFC is consistent, then its predictions about finite calculations will be correct and all is well. But there is always the chance that an inconsistency will be found. Those who worry about this risk can take some comfort in two thoughts: First, many mathematicians have worked with ZFC for many years, and no contradictions have been found yet. And, second, even if a contradiction is found, it will most likely be possible to make small changes to eliminate it without losing any important mathematics in the process.[2]

By contrast, there is a completely different kind of response that reflection on Hilbert's program has inspired. The American mathematician Stephen Simpson (1945–) has described it as follows:[3]

> Gödel's theorem shows that it is impossible to reduce *all* of infinitistic mathematics to finitistic mathematics. There remains the problem of validating as much of infinitistic mathematics as possible. In particular, what *part* of infinitistic mathematics can be reduced to finitistic reasoning?

Significant progress has been made on this project of achieving a partial realization of Hilbert's program. The American mathematician Harvey Friedman (1948–) has defined a formal system, known as WKL_0, in which

---

[2] Something like this view may well be the one that many mathematicians today fall back on when they fail to explain the phenomenon of mathematics in any other way. For example, the view described here is simila in some respects to the view presented in Henle (1991). See also Curry (1951, 1954).

[3] Simpson (1988, 353).

significant parts of infinitary mathematics can be carried out, and he has proven that WKL_0 is a conservative extension of finitary mathematics for finitarily meaningful statements. More recently, the German philosopher Wilfried Sieg (1948–) has given a finitary proof of this conservation result. Theorems provable in WKL_0 include, for example, those that establish many of the most important properties of continuous functions from $\mathbb{R}$ to $\mathbb{R}$. However, WKL_0 does not encompass the study of discontinuous functions. Furthermore, by Gödel's Incompleteness Theorems, we know that in WKL_0 we cannot even prove all of the theorems of PA. In fact, WKL_0 includes only a weak form of the induction axiom of PA. Thus, while this work provides precisely the kind of justification that Hilbert sought, it does so only for a limited part of infinitary mathematics.[4]

This research shows that it is possible to use finitary reasoning to justify a significant part of infinitary mathematics. If one is willing to go beyond finitary reasoning, then it is possible to justify even more of infinitary mathematics. In fact, we have already seen an example of this. Recall, from Chapter 5, Gödel's proof that any theorem of PA that contains neither disjunctions nor existential quantifiers is also a theorem of HA, a formalization of intuitionistic arithmetic. This result should lead an intuitionist to trust proofs in PA, at least with respect to theorems that contain neither disjunctions nor existential quantifiers. In 1936, Gentzen provided another intuitionistic justification for PA by using an extension of the method of mathematical induction called *transfinite induction*.[5] More recently, a number of researchers, including the Japanese mathematician Gaisi Takeuti (1926–), the American mathematician Solomon Feferman (1928–), and Friedman, have extended Gentzen's methods to find intuitionistic justifications for stronger infinitary theories. This work culminated in theorems of Sieg and the German mathematicians Wilfried Buchholz (1948–) and Wolfram Pohlers (1943–) in 1977, and provided an intuitionistic justification for a theory known as $\Pi_1^1 - CA_0$.[6] This theory is stronger than both PA and WKL_0, and allows for certain impredicative definitions of sets of natural numbers. Theorems provable in $\Pi_1^1 - CA_0$ but not WKL_0 include the Bolzano–Weierstrass Theorem and the Cantor–Bendixson Theorem.[7] Thus, al-

---

[4] For more on this research, see Simpson (1999).

[5] Gentzen (1936). See also Troelstra and Schwichtenberg (1996).

[6] A more detailed history of this research can be found in the preface to Buchholz et al. (1981).

[7] The ideas behind the development of analysis in $\Pi_1^1 - CA_0$ can be traced back to Hilbert and Bernays (1934–9); see, in particular, Supplement IV of volume 2. For more information about the strength of the theory $\Pi_1^1 - CA_0$, see Simpson (1999).

though Hilbert's project in the form in which he originally envisaged it is unrealizable, we see that significant partial success can be achieved if we recast it more generally as an attempt to justify classical infinitary reasoning from a point of view that does not accept the completed infinite.

In this chapter, we have sketched how our three different approaches to mathematical foundations respond to, and in the process reshape, the philosophical problems with which we began. Clearly, each approach has led to dashed expectations. Nevertheless, each also has the virtue of all grand ideas, namely that of encouraging and rewarding inventive attempts to transform disappointment into delight. Consequently, we have also tried here to indicate briefly some of the interesting directions of inquiry, both mathematical and philosophical, that these approaches have inspired.

It would take several more volumes to do justice to all the valuable work that rides upon the shoulders of logicism, intuitionism, and the finitist program. In this book, we have sought only to offer the reader a first rung on the ladder.

References

Abel, N. H. 1902: *Mémorial: publié à l'occasion du centenaire de sa naissance.* Dybwad: Kristiana/London: Williams & Norgate.

Auden, W. H. and Kronenberger, L. (eds.) 1966: *The Viking Book of Aphorisms.* New York: Viking Press.

Bell, E. T. 1937: *Men of Mathematics.* New York: Simon and Schuster.

Benacerraf, Paul and Putnam, Hilary (eds.) 1983: *Philosophy of Mathematics: Selected Readings*, 2nd edn. Cambridge: Cambridge University Press.

Bernays, Paul 1922: Über Hilberts Gedanken zur Grundlegung der Arithmetik. *Jahresbericht der Deutschen Mathematiker Vereiningung*, 31, 10–19. Reprinted as "On Hilbert's thoughts concerning the grounding of arithmetic" in Mancosu (1998), op. cit., pp. 215–22.

—— 1923: Reply to the note by Mr. Aloys Müller, "On numbers as signs." Reprinted in Mancosu (1998), op. cit., pp. 223–6.

—— 1930: The philosophy of mathematics and Hilbert's proof theory. Reprinted in Mancosu (1998), op. cit., pp. 234–65.

Boolos, George 1979: *The Unprovability of Consistency: An Essay in Modal Logic.* Cambridge: Cambridge University Press.

—— 1986/87: Saving Frege from contradiction. Reprinted in, Boolos (1998), op. cit., pp. 171–82.

—— 1989: A new proof of the Gödel Incompleteness Theorem. *Notices of the American Mathematical Society*, 36(4), 388–90; reprinted in Boolos (1998), op. cit., pp. 383–8.

—— 1998: *Logic, Logic, and Logic.* Cambridge, Mass.: Harvard University Press.

Buchholz, Wilfried, Feferman, Solomon, Pohlers, Wolfram, and Sieg, Wilfried 1981: *Iterated Inductive Definitions and Subsystems of Analysis: Recent Proof-Theoretical Studies.* Lecture Notes in Mathematics, vol. 897. Berlin: Springer-Verlag.

Curry, Haskell B. 1951: *Outlines of a Formalist Philosophy of Mathematics.* Amsterdam: North-Holland.

—— 1954: Remarks on the definition and nature of mathematics. *Dialectica*, 8, 228–33. Reprinted in Benacerraf and Putnam (1983), op. cit., pp. 202–6.

Cutland, N. J. 1980: *Computability: An Introduction to Recursive Function Theory.* Cambridge: Cambridge University Press.

Dauben, Joseph 1979: *Georg Cantor: His Mathematics and Philosophy of the Infinite*. Cambridge, Mass.: Harvard University Press.

Dedekind, Richard 1888: *Was sind und was sollen die Zahlen?* In *Essays on the Theory of Numbers*, transl. Wooster Woodruff Beman. New York: Dover, 1963, pp. 31–115.

Demopoulos, William (ed.) 1995: *Frege's Philosophy of Mathematics*. Cambridge, Mass.: Harvard University Press.

Dummett, Michael 1963: The philosophical significance of Gödel's Theorem. Reprinted in Michael Dummett 1978: *Truth and Other Enigmas*. Cambridge, Mass.: Harvard University Press, pp. 186–201.

—— 1978: The philosophical basis of intuitionistic logic. In *Truth and Other Enigmas*. Cambridge, Mass.: Harvard University Press, pp. 215–47.

—— 1991: *Frege: Philosophy of Mathematics*. Cambridge, Mass.: Harvard University Press.

—— 1994: What is mathematics about? In Alexander George (ed.), *Mathematics and Mind*. Oxford: Oxford University Press, pp. 11–26.

—— 2000: *Elements of Intuitionism*, 2nd edn. Oxford: Oxford University Press.

Einstein, Albert 1983: *Sidelights on Relativity*. New York: Dover.

Enderton, Herbert B. 2001: *A Mathematical Introduction to Logic*, 2nd edn. San Diego: Harcourt/Academic Press.

Ewald, William B. 1996: *From Kant to Hilbert: A Source Book in the Foundations of Mathematics*. Oxford: Clarendon Press.

Feferman, Solomon, Dawson, John W. Jr., Kleene, Stephen C., Moore, Gregory H., Solovay, Robert M., and van Heijenoort, Jean (eds.), *Kurt Gödel: Collected Works*, vol. 1. Oxford: Oxford University Press, 1986.

Frege, Gottlob 1884: *Die Grundlagen der Arithmetik*. Translated by J. L. Austin as *The Foundations of Arithmetic*. Evanston, Ill.: Northwestern University Press, 1978.

—— 1892/1903: *Die Grundgesetze der Arithmetik* (vol. 1, 1892; vol. 2, 1903). Partially translated and edited in M. Furth 1964: *The Basic Laws of Arithmetic: Exposition of the System*. Berkeley: University of California Press.

Geach, Peter and Black, Max (eds.) 1980: *Translations from the Philosophical Writings of Gottlob Frege*, 3rd edn. Oxford: Blackwell.

Gentzen, Gerhard 1936: The consistency of elementary number theory. Reprinted in M. E. Szabo (ed.), *The Collected Papers of Gerhard Gentzen*. Amsterdam: North-Holland, 1969, pp. 132–70.

George, Alexander and Velleman, Daniel J. 1998: Two conceptions of natural number. In H. G. Dales and G. Oliveri (eds.), *Truth in Mathematics*. Oxford: Oxford University Press, pp. 311–27.

—— and —— 2000: Leveling the playing field between mind and machine: reply to McCall. *The Journal of Philosophy*, XCVII(8), 456–61.

Gödel, Kurt 1931: On formally undecidable propositions of *Principia mathematica* and related systems I. Reprinted in Feferman et al. (1986), op. cit., pp. 145–95.

—— 1933: On intuitionistic arithmetic and number theory. Reprinted in Feferman et al. (1986), op. cit., pp. 282–95.

Hale, Bob and Wright, Crispin 2001: *The Reason's Proper Study: Essays towards a Neo-Fregean Philosophy of Mathematics*. Oxford: Clarendon Press.

Hallett, Michael 1984: *Cantorian Set Theory and Limitation of Size*. Oxford: Clarendon Press.

Hardy, G. H. 1967: *A Mathematician's Apology*. Cambridge: Cambridge University Press.

Heck, Richard G. Jr. 1995: The development of arithmetic in Frege's *Grundgesetze der Arithmetik*. Reprinted in Demopoulos (1995), op. cit., pp. 257–94.

van Heijenoort, Jean (ed.) 1967: *From Frege to Gödel: A Source Book in Mathematical Logic, 1879–1931*. Cambridge, Mass.: Harvard University Press.

Henle, J. M. 1991: The happy formalist. *The Mathematical Intelligencer*, 13(1), 12–18.

Herbrand, Jacques 1931: On the consistency of arithmetic. Reprinted in van Heijenoort (1967), op. cit., pp. 618–28.

Hermes, Hans, Kambartel, Friedrich, and Kaulbach, Friedrich (eds.) 1979: *Posthumous Writings*. Oxford: Basil Blackwell.

Heyting, Arend 1976: *Intuitionism: An Introduction*. Amsterdam: North-Holland.

Hilbert, David 1918: Axiomatic thought. Translated in vol. 2 of Ewald (1996), op. cit., pp. 1107–15.

—— 1922: The new grounding of mathematics: first report. Translated in vol. 2 of Ewald (1996), op. cit., pp. 1117–34.

—— 1925: On the infinite. Reprinted in van Heijenoort (1967), op. cit., pp. 367–92.

—— 1927: The foundations of mathematics. Reprinted in van Heijenoort (1967), op. cit., pp. 464–79.

—— and Bernays, Paul 1934–9: *Grundlagen der Mathematik*, 2 vols. Berlin: Julius Springer.

Hume, David 1748: *An Enquiry Concerning Human Understanding*. Edited by Tom L. Beauchamp. Oxford: Oxford University Press. See §7, "Of the idea of necessary connexion."

Jarden, Dov 1953: A simple proof that a power of an irrational number to an irrational exponent may be rational. *Scripta Mathematica*, XIX, 229.

Kant, Immanuel 1783: *Prolegomena to Any Future Metaphysics*, translated by L. W. Beck. Indianapolis, Ind.: Bobbs-Merrill, 1950.

Kreisel, Georg 1983: Hilbert's programme. In Benacerraf and Putnam (1983), op. cit., pp. 207–38.

Mancosu, Paolo 1998: *From Brouwer to Hilbert: The Debate on the Foundations of Mathematics in the 1920s*. Oxford: Oxford University Press.

Mill, John Stuart 1843: *A System of Logic*. London: Longman, Green.

Moore, Gregory 1982: *Zermelo's Axiom of Choice: Its Origins, Development, and Influence*. New York: Springer-Verlag.

224   *References*

Moritz, Robert Edouard 1958: *On Mathematics and Mathematicians*. New York: Dover.

Peano, Giuseppe 1889: The principles of arithmetic, presented by a new method. Translated in van Heijenoort (1967), op. cit., pp. 83–97.

Poincaré, Henri 1905: Les mathématiques et la logique. Translated in vol. 2 of Ewald (1996), op. cit., pp. 1021–71.

Quine, W. V. 1962: Paradox. *Scientific American*, 206. Reprinted as "The ways of paradox" in *The Ways of Paradox and Other Essays*, revised and enlarged edn. Cambridge, Mass.: Harvard University Press, 1976, pp. 1–18.

—— 1969: *Set Theory and Its Logic*. Cambridge, Mass.: Harvard University Press.

—— 1970: *Philosophy of Logic*. Englewood Cliffs, NJ: Prentice-Hall.

Reid, Constance 1970: *Hilbert*. Berlin: Springer-Verlag.

Russell, Bertrand 1905: On some difficulties in the theory of transfinite numbers and order types. Reprinted in Douglas Lackey (ed.), *Essays in Analysis*. New York: George Braziller, 1973.

—— 1956: *Portraits from Memory and Other Essays*. New York: Simon and Schuster.

Salmon, Nathan and Soames, Scott (eds.) 1988: *Propositions and Attitudes*. Oxford: Oxford University Press.

Sieg, Wilfried 1984: Foundations for analysis and proof theory. *Synthese*, 60(2), 162.

Simpson, Stephen G. 1988: Partial realizations of Hilbert's program. *Journal of Symbolic Logic*, 53, 349–63.

—— 1999: *Subsystems of Second Order Arithmetic*. Berlin: Springer-Verlag.

Smullyan, Raymond 1978: *What Is the Name of This Book?* Englewood Cliffs, NJ: Prentice-Hall.

—— 1982: *The Lady or the Tiger?* New York: Alfred A. Knopf.

Strichartz, Robert S. 1995: *The Way of Analysis*. Boston: Jones and Bartlett.

Tarski, Alfred 1933: The concept of truth in formalized languages. Reprinted in his *Logic, Semantics, Metamathematics: Papers from 1923 to 1938*, 2nd edn. Indianapolis, Ind.: Hackett, 1983, pp. 152–278.

Troelstra, A. S. and van Dalen, D. 1988: *Constructivism in Mathematics: An Introduction*. Amsterdam: North-Holland.

Troelstra, A. S. and Schwichtenberg, H. 1996: *Basic Proof Theory*. Cambridge: Cambridge University Press.

Velleman, Daniel J. 1993: Constructivism liberalized. *The Philosophical Review*, 102, 59–84.

Weinberg, Steven 1986: In "Mathematics: the unifying thread in science" (transcript of a symposium). *Notices of the American Mathematical Society*, XXXIII (October), 716–33. See pp. 725–8.

Weyl, Hermann 1925–7: The current epistemological situation in mathematics. Reprinted in Mancosu (1998), op. cit., pp. 123–42.

Wittgenstein, Ludwig 1978: *Remarks on the Foundations of Mathematics*, edited by G. H. von Wright, R. Rhees and G. E. M. Anscombe; translated by G. E. M. Anscombe. Cambridge, Mass.: The MIT Press, §121.

Index

Also Available from Blackwell Publishers:

Deduction
Introductory Symbolic Logic
Second Edition
Daniel Bonevac
0631227105 hardback
063122713X paperback 2002

The Blackwell Guide to Philosophical Logic
Edited by Lou Goble
Series: Blackwell Philosophy Guides
0631206922 hardback
0631206930 paperback 2001

An Introduction to Philosophical Logic
Third Edition
A. C. Grayling
0631206558 hardback
0631199829 paperback 1997

Languages of Logic
An Introduction to Formal Logic
Second Edition
Samuel Guttenplan
155786988X paperback 1997

A Companion to Philosophical Logic
Edited by Dale Jacquette
Series: Blackwell Companions to Philosophy
0631216715 hardback 2002

Philosophy of Logic
An Anthology
Edited by Dale Jacquette
Series: Blackwell Philosophy Anthologies
063121867X hardback
0631218688 paperback 2001

Philosophy of Mathematics
An Anthology
Edited by Dale Jacquette
Series: Blackwell Philosophy Anthologies
0631218696 hardback
063121870X paperback 2001

Meaning and Argument
An Introduction to Logic through Language
Edited by Ernest Lepore
0631205829 paperback 2000

Counterfactuals
David Lewis
0631224955 hardback
0631224254 paperback 2000

Analytic Philosophy
An Anthology
Edited by A. P. Martinich and David Sosa
Series: Blackwell Philosophy Anthologies
0631216464 hardback
0631216472 paperback 2001

A Companion to Analytic Philosophy
Edited by A. P. Martinich and David Sosa
Series: Blackwell Companions to Philosophy
0631214151 hardback 2001

Logical Forms
An Introduction to Philosophical Logic
Second Edition
Mark Sainsbury
0631216782 hardback
0631216790 paperback 2000

To receive FREE e-mail updates on Blackwell philosophy titles register for SELECT at
http://select.blackwellpublishers.co.uk

Blackwell books are available from your local bookstore, or can be ordered direct from our website www.blackwellpublishers.co.uk

To request a free examination copy* e-mail exam@blackwellpub.com (The Americas & Canada) or inspection@marston.co.uk (Europe and ROW).

* examination copies available for paperback titles only